INSTANT PHYSIOLOGY

Neil R. Borley

FRCS, FRCS Ed
Clinical Tutor in Surgery
University of Oxford
Oxford Radcliffe Hospital
Oxford

Vinod Achan

MA, MRCP
Cardiovascular Research Fellow
Stanford University School of
Medicine
Stanford
California

b
**Blackwell
Science**

© 2000
Blackwell Science Ltd
Editorial Offices:
Osney Mead, Oxford OX2 0EL
25 John Street, London WC1N 2BL
23 Ainslie Place, Edinburgh EH3 6AJ
350 Main Street, Malden
  MA 02148-5018, USA
54 University Street, Carlton
  Victoria 3053, Australia
10, rue Casimir Delavigne
  75006 Paris, France

Other Editorial Offices:
Blackwell Wissenschafts-Verlag GmbH
Kurfürstendamm 57
10707 Berlin, Germany

Blackwell Science KK
MG Kodenmacho Building
7–10 Kodenmacho Nihombashi
Chuo-ku, Tokyo 104, Japan

The right of the Authors to be
identified as the Authors of this Work
has been asserted in accordance
with the Copyright, Designs and
Patents Act 1988.

First published 2000

Set by Graphicraft Limited, Hong Kong
Printed and bound in Great Britain by
MPG Books Ltd, Bodmin, Cornwall

DISTRIBUTORS

Marston Book Services Ltd
PO Box 269
Abingdon, Oxon OX14 4YN
(*Orders*: Tel: 01235 465500
        Fax: 01235 465555)

USA
Blackwell Science, Inc.
Commerce Place
350 Main Street
Malden, MA 02148-5018
(*Orders*: Tel: 800 759 6102
          781 388 8250
    Fax: 781 388 8255)

Canada
Login Brothers Book Company
324 Saulteaux Crescent
Winnipeg, Manitoba R3J 3T2
(*Orders*: Tel: 204 837 2987)

Australia
Blackwell Science Pty Ltd
54 University Street
Carlton, Victoria 3053
(*Orders*: Tel: 3 9347 0300
    Fax: 3 9347 5001)

A catalogue record for this title
is available from the British Library

ISBN 0-632-05004-7

Library of Congress
Cataloging-in-publication Data

Borley, Neil R.
    Instant physiology/
    Neil R. Borley, Vinod Achan.
        p.   cm.
    ISBN 0-632-05004-7
    1. Human physiology Outlines, syllabi, etc.
    I. Achan, Vinod.   II. Title.
QP41.B757   2000
612′.002′02—dc21   99-29385
                                      CIP

For further information on
Blackwell Science, visit our website:
www.blackwell-science.com

To Alison
and to Kalpa, Ravi and Maya

CONTENTS

PREFACE

Those of us preparing for the basic medical science, MRCP, MRCS and other examinations, do so with less time to spare and under increasing pressures of work. Revision typically takes place during brief windows of opportunity in a relatively fragmented fashion. At the same time, the emphasis on basic sciences as they relate to clinical practice has increased in these examinations. In particular, it is widely recognized amongst clinicians that an understanding of physiology is essential to the understanding of disease.

Instant Physiology was conceived to allow easy access to physiological principles as they relate to medical training by packaging important subject areas into manageable portions. Key topics are (as far as possible) presented in a 'one topic per two-page spread' format. This allows the reader to view the information on a selected topic at a single glance without the need for repeated cross-referencing between pages. Diagrams, tables and lists are widely used to reduce the amount of information presented as text, while clinical and pharmacological boxes are included to emphasize the clinical relevance of topics. Molecular aspects of physiology have also been discussed where appropriate to reflect the increasing contribution of detail at this level.

The topics we have chosen to cover here will inevitably be a compromise; some subject areas are dealt with in less detail than might be expected or omitted altogether. Above all, *Instant Physiology* is not a text book. It is designed to be used for 'dipping in' and key topic revision rather than to be read chapter by chapter as one would a conventional text. In many ways *Instant Physiology* represents the book we would like to have had next to us when revising for our own examinations. We would welcome any comments or suggestions for future improvements; these can be sent by email to authors@blacksci.co.uk.

LIST OF ABBREVIATIONS

2,3-DPG	2,3-diphosphoglycerate
5HT	5-hydroxytryptamine
A	angiotensin (receptor)
AI	angiotensin I
AII	angiotensin II
AIR	angiotensin I receptor
AIIR	angiotensin II receptor
ABP	arterial blood pressure
AC	adenylyl cyclase
ACE	angiotensin-converting enzyme
ACh	acetylcholine
AChE	acetylcholinesterase
AChR	acetylcholine receptor
ACTH	adrenocorticotrophic hormone
AD	autosomal dominant
ADH	antidiuretic hormone
ADMA	asymmetrical dimethylarginine
ADP	adenosine diphosphate
ALP	alkaline phosphatase
AMP	adenosine monophosphate
ANP	atrial natriuretic peptide
ANP-A, ANP-B	atrial natriuretic peptide A, B
AoP	aortic pressure
AoV	aortic valve
AP	action potential
AR	aortic regurgitation
AR	autosomal recessive
ARF	acute renal failure
AS	aortic stenosis
ASD	atrial septal defect
AST	aspartate aminotransferase
ATIII	antithrombin III
ATN	acute tubular necrosis
ATP	adenosine triphosphate

ATPase	adenosine triphosphatase
a-v	arteriovenous
AV	atrioventricular
BF	blood flow
BMI	body mass index
BNP	brain natriuretic peptide
BP	blood pressure
CaMK	calmodulin-dependent kinase
cAMP	cyclic adenosine monophosphate
CBG	cortisol-binding globulin
CCD	cortical collecting duct
CCK	cholecystokinin
CF	cystic fibrosis
cGMP	cyclic guanosine monophosphate
CK	creatine kinase
CNP	C-type natriuretic peptide
CNS	central nervous system
CO	cardiac output
COHb	carboxyhaemoglobin
COMT	catecholamine O-methyl transferase
COX	cyclooxygenase
CRE	cAMP-responsive element
CREB	cAMP-responsive element-binding protein
CRF	chronic renal failure
CSF	cerebrospinal fluid
CVA	cerebrovascular accident
CVP	central venous pressure
DA	ductus arteriosus
DAG	1,2-diacylglycerol
DCT	distal convoluted tubule
DI	diabetes insipidus
DIT	di-iodotyrosine
DKA	diabetic ketoacidosis
Do_2	oxygen delivery
E	endothelin (receptor)
E_m	membrane potential

E_{rev}	Nernst potential
EBV	Epstein–Barr virus
EC	endothelial cell
ECE	endothelin-converting enzyme
ECF	extracellular fluid
ECG	electrocardiogram
ECL	enterochromaffin-like cell
EDV	end-diastolic volume
EF	ejection fraction
EGFR	epidermal growth factor receptor
ENaC	epithelial Na channel
Enk	enkephalin
eNOS	endothelial NOS
Epo	erythropoietin
EPP	end-plate potential
ERV	expiratory reserve volume
ESR	erythrocyte sedimentation rate
ESV	end-systolic volume
ET	endothelin
ETA	endothelin type A receptor
ETB	endothelin type B receptor
FDPs	fibrin degradation products
FEV	forced expiratory volume
FEV_1	forced expiratory volume in 1 s
FFA	free fatty acids
FO	foramen ovale
FRC	functional residual capacity
FRV	functional residual volume
FSH	follicle-stimulating hormone
FVC	forced vital capacity
γ-GT	γ-glutamyl transferase
G6PD	glucose-6-phosphate dehydrogenase
GABA	γ-aminobutyric acid
GB	gallbladder
GC	guanylate cyclase
GCSF	granulocyte colony-stimulating factor
GDP	guanosine diphosphate

GFR	glomerular filtration rate
GH	growth hormone
GHR	growth hormone receptor
GHRH	growth hormone-releasing hormone
GI	gastrointestinal
GIP	gastric inhibitory peptide
GIT	gastrointestinal tract
GMCSF	granulocyte/macrophage colony-stimulating factor
GT	glucuronyl transferase
GTN	glyceryl trinitrate
GTP	guanosine triphosphate
GTPase	guanosine triphosphatase
Hb	haemoglobin
HDL	high-density lipoprotein
$HeSO_4$	heparin sulphate
HR	heart rate
HUS	haemolytic–uraemic syndrome
ICBP	intracellular binding protein
ICF	intracellular fluid
IDL	intermediate-density lipoprotein
IF	intrinsic factor
IFNγ	interferon-γ
Ig	immunoglobulin
IGF	insulin-like growth factor
IGF-1	insulin-like growth factor 1
IGF-2	insulin-like growth factor 2
IL	interleukin
iNOS	inducible NOS
IP_3	inositol 1,4,5-triphosphate
IRV	inspiratory reserve volume
ITP	idiopathic thrombocytopenic purpura
iu	international unit
JGA	juxtaglomerular apparatus
LA	left atrium (or atrial)
LAD	left anterior descending

LAP	left atrial pressure
LCX	left circumflex artery
LDH	lactate dehydrogenase
LDL	low-density lipoprotein
LH	luteinizing hormone
LOS	lower oesophageal sphincter
LV	left ventricle (or ventricular)
LVP	left ventricular pressure
LVED	left ventricular end-diastolic
LVOTO	left ventricular outflow tract obstruction
M	muscarinic (receptor)
MAO	monoamine oxidase
MAP	mean arterial pressure
MB	myocardial type (or band)
MCH	mean corpuscular haemoglobin
MCHC	mean corpuscular haemoglobin concentration
MCV	mean cell volume
MEPP	miniature end-plate potential
MG	myasthenia gravis
MI	myocardial infarction
MIT	mono-iodotyrosine
MLF	medial longitudinal fasciculus
MR	mitral regurgitation
MRF	mesencephalic reticular formation
mRNA	messenger RNA
MUGA	multiple uptake gated acquisition
MV	mitral valve
N	nicotinic (receptor)
NE	norepinephrine
NMJ	neuromuscular junction
nNOS	neuronal NOS
NO	nitric oxide
NOS	nitric oxide synthetase
NPR-A	natriuretic peptide receptor A
NPR-B	natriuretic peptide receptor B
OxLDL	oxidized low-density lipoprotein

P_a	arterial partial pressure
P_A	alveolar partial pressure
PA	pulmonary artery
$P_A\text{CO}_2$	alveolar CO_2 partial pressure
PAH	*para*-aminohippuric acid
PAP	pulmonary arterial pressure
PBC	primary biliary cirrhosis
PC and PS	protein C and S
PC	phospholipase C
PCT	proximal convoluted tubule
PCV	packed cell volume
PCWP	pulmonary capillary wedge pressure
PDA	patent ductus arteriosus
PDGF	platelet-derived growth factor
$P_E\text{CO}_2$	mean end-tidal CO_2 partial pressure
PF3	platelet factor 3
PGE	prostaglandin E
P_i	inspired partial pressure
PIP_2	phosphatidylinositol 4,5-biphosphate
PKA	protein kinase A
PKC	protein kinase C
PP	pancreatic polypeptide
PPRF	paramedial prepontine reticular formation
PR	pulse rate
PTH	parathyroid hormone
PTHrP	PTH-related peptide
P_v	venous partial pressure
PV	pulmonary vein/venous
PVP	pulmonary venous pressure
PVR	peripheral vascular resistance
RA	right atrium (or atrial)
RAP	right atrial pressure
RAS	renin–angiotensin system
RBC	red blood cell
RBCC	red blood cell count
RBF	renal blood flow
RCA	right coronary artery
RPF	renal plasma flow

RR	respiratory rate
RTA	renal tubular acidosis
RV	residual volume
RV	right ventricle (or ventricular)
RVOTO	right ventricular outflow tract obstruction
SA	sinoatrial
SAH	subarachnoid haemorrhage
SC	superior colliculus
SCD	sickle cell disease
SIADH	syndrome of inappropriate antidiuretic hormone secretion
SLE	systemic lupus erythematosus
SMC	smooth muscle cell
SMR	standardized mortality rate
SN	sinus node
SR	sarcoplasmic reticulum
SV	stroke volume
SVC	superior vena cava
SVR	systemic vascular resistance
T	thrombin
T_3	triiodothyronine
T_4	thyroxine
TAL	thin ascending limb
TAPVD	total anomalous pulmonary venous drainage
TBG	thyroxine-binding globulin
TBW	total body water
TDL	thin descending limb
TF	tissue factor
TFTs	thyroid function tests
TGFβ	transforming growth factor-β
THAL	thick ascending limb
TIBC	total iron-binding capacity
TLC	total lung capacity
TM	thrombomodulin
TNFα	tumour necrosis factor-α
TOF	tetralogy of Fallot
TPA	tissue plasminogen activator

TPN	total parenteral nutrition
TSH	thyroid-stimulating hormone
TTP	thrombotic thrombocytic purpura
TV	tricuspid valve
uNa$^+$	urinary sodium concentration
V_A	alveolar ventilation
VC	vena cava
VC	vital capacity
VIP	vasoactive intestinal peptide
VLDL	very low-density lipoprotein
VMC	vasomotor centre
V_{O_2}	oxygen consumption
VSD	ventricular septal defect
V_T	tidal volume
vWF	von Willebrand factor
WCC	white cell count
Xn	vagus nerve

1: CELLULAR PHYSIOLOGY

Transmembrane solute transport

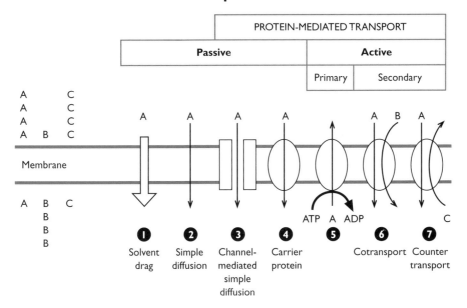

Fig. 1
A, B and C are different molecules.

Ion channels

Ion channels are protein tunnels spanning the cell membrane. Channel opening results in a current of the order of a few picoamps generated by the flow of highly specific ions.

Potassium channels

(a) Outward or delayed rectifier K+ channel (Kv)
Activated by membrane depolarization

Produces an outward K+ current

Responsible for the repolarization of the cardiac action potential

(b) ATP-sensitive K+ channel (K+-ATP)
Accelerates repolarization

Shortens the cardiac action potential

Prostacyclin, vasoactive intestinal peptide (VIP), nitric oxide (NO) and adenosine act in part via K+-ATP opening

K+-ATP channels open during ischaemia in response to a fall in intracellular ATP, acidosis, a rise in ADP and GDP, and the accumulation of extracellular adenosine

Antianginal (nicorandil) and vasodilator agents (diazoxide and minoxidil) act via myocyte K+-ATP opening. Sulphonylureas such as glibenclamide are selective K+-ATP blockers

(c) G-protein-activated K+ channel (K-ACh)
Opened by vagally secreted acetylcholine (ACh)

Decreases spontaneous depolarization in the sinus node

Slows atrioventricular (AV) node conduction, underlying the vagal slowing of heart rate

(d) Inwardly rectifying K+ channel
Opens at very negative potentials (less than -40 mV)

Shows a reduced K+ conductance at positive membrane potentials (opposite to normal outward rectification seen in delayed rectifier channels)

K+-ATP and K+-ACh display some inward rectification

Calcium channels

(a) L-type Ca2+ channel (long lasting)
High voltage activated

Expressed in vascular and cardiac tissue

Generates a slow inward current

Blocked by dihydropyridines (nifedipine, amlodipine)

(b) T-type Ca2+ channel (transient)
Low voltage activated

Rapidly inactivated

High expression in the sinus node—possible role in pacemaking

Blocked by verapamil, diltiazem

(c) N-, P-, Q- and R-type Ca2+ channels
Found in neuronal cells

Ion channel disorders

Disorder		Channel	Clinical notes
Bartter's syndrome	AR	Bumetanide-sensitive Na+K+Cl– cotransporter (NKCC2)	*Hypokalaemia, alkalosis, renal salt wasting, hypotension, hyperreninaemia, hyperaldosteronism*
Liddle's syndrome (hereditary hypertension)	AR	ENaC (epithelial Na channel)	
Hyperkalaemic periodic paralysis	AD	Skeletal muscle Na channel	
Hypokalaemic periodic paralysis	AD	L-type Ca^{2+} channel	
Becker's generalized myotonia	AR	Skeletal muscle Cl channel	
Long QT syndrome	AD	Type 1, KVLQT1 (cardiac K+ channel) Type 2, HERG (cardiac K+ channel) Type 3, SCN5A (cardiac Na+ channel)	*Characterized by prolonged and abnormal ventricular repolarization and risk of life-threatening arrhythmias (particularly torsades de pointes)*

AD, autosomal dominant; AR, autosomal recessive.

Cystic fibrosis (CF)

- The CF transmembrane conductance regulator (*CFTR*) gene is defective in CF.
- CFTR is a cAMP-regulated Cl channel found in the apical membrane of epithelial cells.
- CFTR also downregulates Na absorption via the amiloride-sensitive ENaC.
- Reduced Cl transport is thought to reduce Cl and water secretion into the airway lumen.

I

Ion ATPases

Na⁺/K⁺-ATPase

The chemical energy of ATP hydrolysis is used to extrude three Na^+ ions for every two K^+ ions entering the cell and every ATP molecule hydrolysed.

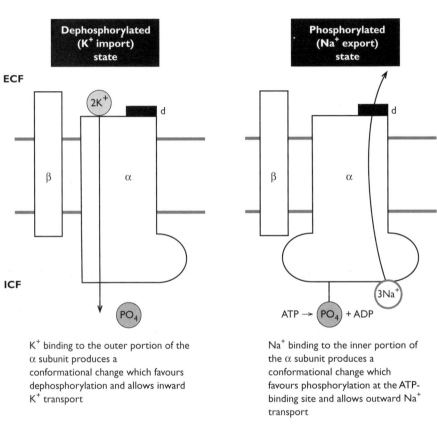

Fig. 2 Na⁺/K⁺-ATPase function
d, digoxin-binding site; ECF, extracellular fluid; ICF, intracellular fluid.

There is a net export of one third of a positive charge per Na^+ ion transported. Intracellular Na^+ is the substrate of the pump and a rise in intracellular Na^+ concentration favours Na^+/K^+ exchange.

Na⁺/K⁺-ATPase maintains intracellular and extracellular Na^+ and K^+ concentrations and is thus responsible for maintaining the resting membrane potential. The active transport of Na^+ is also coupled to the transport of other substances (secondary active transport, counter transport and cotransport).

Magnesium is a cofactor of Na^+/K^+-ATPase and thus helps to maintain intracellular K^+.

Digoxin is an Na^+/K^+-ATPase inhibitor and thus produces a rise in intracellular Na^+ as well as a fall in intracellular K^+.

Other ATPases

Gastric H^+/K^+-ATPase

- Responsible for hydrogen ion secretion.
- *Antigen recognized by parietal cell autoantibodies in pernicious anaemia.*

Ca^{2+}/Mg^{2+}-ATPase

- Actively pumps Ca^{2+} into the sarcoplasmic reticulum during muscular relaxation (see 'Excitation–contraction coupling', p. 172).

H^+-ATPase

- Responsible for acid secretion in the distal convoluted tubule and collecting duct of the kidney.
- *A deficiency of this active proton pump (as in Sjögren's syndrome) results in distal (type 1) renal tubular acidosis* (see 'Renal', p. 97).

Resting membrane potential (E_m)

The Nernst potential (E_{rev}) for an ion is the point at which chemical and electrical driving forces across the cell membrane (occurring in opposite directions) are in equilibrium. At this potential, there is no net flow of that specific ion.

Ion	Extracellular concentration (mmol L^{-1})	Intracellular concentration (mmol L^{-1})	Nernst potential (mV)
Na$^+$	142	10	+70 (E_{Na})
K$^+$	4	155	−98 (E_K)
Ca^{2+}	2.5	0.0001	+150 (E_{Ca})
Cl$^-$	101	5–30	+30 to −65 (E_{Cl})

Under physiological conditions, Na$^+$, Ca^{2+} and Cl$^-$ flow into cells to depolarize the cell towards E_{Na}, E_{Ca} and E_{Cl} respectively. Similarly, K$^+$ flows out of the cell to repolarize the cell towards E_K. E_m depends on the distribution of Na$^+$, Ca^{2+}, Cl$^-$ and K$^+$ as well as membrane permeability to these ions.

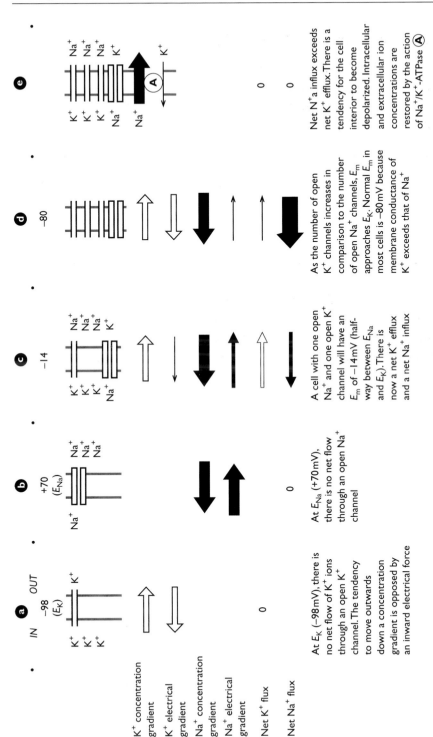

Fig. 3

K⁺ concentration gradient

K⁺ electrical gradient

Na⁺ concentration gradient

Na⁺ electrical gradient

Net K⁺ flux

Net Na⁺ flux

a OUT IN −98 (E_K)

At E_K (−98 mV), there is no net flow of K⁺ ions through an open K⁺ channel. The tendency to move outwards down a concentration gradient is opposed by an inward electrical force

Net K⁺ flux 0

b +70 (E_{Na})

At E_{Na} (+70 mV), there is no net flow through an open Na⁺ channel

Net Na⁺ flux 0

c −14

A cell with one open Na⁺ and one open K⁺ channel will have an E_m of −14 mV (half-way between E_{Na} and E_K). There is now a net K⁺ efflux and a net Na⁺ influx

d −80

As the number of open K⁺ channels increases in comparison to the number of open Na⁺ channels, E_m approaches E_K. Normal E_m in most cells is −80 mV because membrane conductance of K⁺ exceeds that of Na⁺

e

Net Na influx exceeds net K⁺ efflux. There is a tendency for the cell interior to become depolarized. Intracellular and extracellular ion concentrations are restored by the action of Na⁺/K⁺-ATPase Ⓐ

Net K⁺ flux 0

Net Na⁺ flux 0

9

Action potential

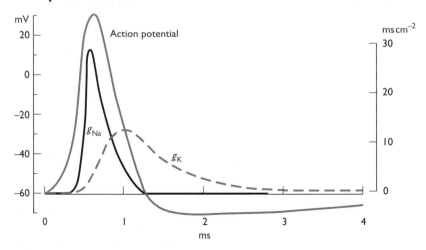

Fig. 4 Axonal action potential
(From Schmidt, R.F. & Thews, G. (eds) (1983) *Human Physiology*. Springer-Verlag, Berlin.)

The action potential is an all or nothing event triggered by the arrival of a depolarizing stimulus when Na^+ influx (g_{Na}) exceeds K^+ efflux (g_K).

Depolarization	When a critical threshold (−55 mV) is reached, all voltage-gated Na^+ channels open, causing E_m to approach E_{Na} (+55 mV) rapidly
Repolarization	A delayed voltage-dependent Na^+ channel inactivation and K^+ channel activation causes E_m to fall, exceeding the resting potential briefly (hyperpolarization) before returning to the starting point

The action potential is followed by an absolute and then relative refractory period.

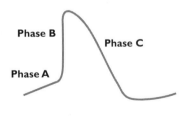

Non-nodal (Purkinje cell)

Phase B

Phase C

Phase A

Phase D

−70 mV

Phase E

Nodal (SA node, AV node)

Phase B

Phase C

Phase A

Resting membrane potential (E_m): −90 mV
Threshold potential: −70 to −60 mV

A Upstroke due to fast Na^+ influx causes depolarization to approximately +15 mV
B E_m repolarization to approximately 0 mV as Na^+ influx stops
C Plateau phase due to voltage-sensitive Ca^{2+} influx and K^+ efflux
D Repolarization due to outward K^+ current (via delayed rectifier K^+ channels)
E Slow upward drift of E_m

Resting membrane potential (E_m): −60 mV
Na^+ channels absent

A Prepotential due to residual Ca^{2+} influx
B Depolarization due to Ca^{2+} influx (via T-type Ca^{2+} channels)
C Repolarization due to outward K^+ current

Fig. 5 Cardiac action potential
SA, sinoatrial; AV, atrioventricular.

Second messenger pathways

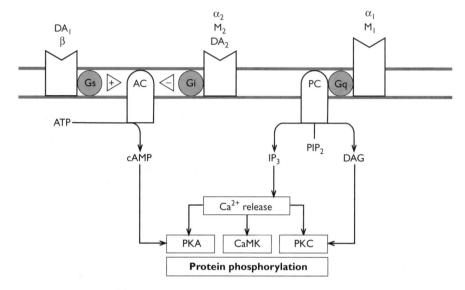

Fig. 6
AC, adenyl cyclase; Ca²⁺; CaMK, calmodulin-dependent kinase; DAG, 1,2-diacylglycerol;
Gs, Gi, Gq, G proteins; IP_3, inositol 1,4,5-triphosphate; PC, phospholipase C; PIP_2,
phosphatidylinositol 4,5-biphosphate; PKA, protein kinase A; PKC, protein kinase C.
Receptors: DA, dopamine; M, muscarinic.

cAMP pathway	IP_3 pathway
Activated β_1 and α_2 adrenergic receptors, for example, act via Gs or Gi proteins to stimulate or inhibit AC respectively	Activated α_1 adrenergic receptors, for example, act via G proteins to stimulate PC
AC induces cAMP synthesis	PC cleaves phosphoinositide to give IP_3 and DAG
cAMP stimulates target gene expression (tyrosine hydroxylase, somatostatin) via: 1 PKA induction	IP_3 mobilizes Ca^{2+} from intracellular stores Ca^{2+} and DAG activate calmodulin kinases and PKC
2 phosphorylation of transcription factors (cAMP-responsive element (CRE)-binding protein, CREB)	These in turn phosphorylate a number of important proteins (epidermal growth factor receptor (EGFR), glycogen synthase)

Notes:
Ca may modulate CREB activity via calmodulin kinases but also induces target gene expression via the
cAMP pathway.
Other second messengers include cGMP (atrial natriuretic peptide (ANP), NO, phototransduction).

Use of second messenger pathways by various agonists

Agonist	cAMP raised	cAMP reduced	IP₃/DAG
ACh		M_2	M_1
Epinephrine	β_1	α_2	α_1
Dopamine	DA_1	DA_2	
ADH	VP_2		VP_1
Histamine	H_2		H_1
Adenosine	A_2	A_1	
Other	TSH	Somatostatin	Gastrin
	LH	AII	CCK
	FSH	5HT	GABA

ADH, antidiuretic hormone; CCK, cholecystokinin; FSH, follicle-stimulating hormone; GABA, γ-aminobutyric acid; 5HT, 5-hydroxytryptamine; LH, luteinizing hormone; TSH, thyroid-stimulating hormone; VP, vasopressin.

G proteins

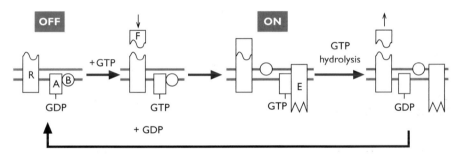

Fig. 7 G-protein function
E, effector molecule; F, first messenger; R, receptor. G protein: A, α subunit; B, β and γ subunits.

G proteins consist of three subunits (A, B, C).
1 In the resting state, GDP is bound to the A subunit which is a GTPase.
2 On hormone binding, GDP is displaced by GTP which activates the G protein.
3 The A subunit and BC complex dissociate to interact with effectors.
4 GTP is then rapidly hydrolysed to GDP.

> G-protein abnormalities are implicated in human disease:
> 1 continued Gs activation is a pathophysiological mechanism in acromegaly, McCune–Albright syndrome and *Vibrio cholerae* infection;
> 2 the oncogene *ras* encodes p21 which is a G protein;
> 3 Gs activity is reduced by 50% in pseudohypoparathyroidism.

2: CARDIOVASCULAR PHYSIOLOGY

2

Basic principles – overview

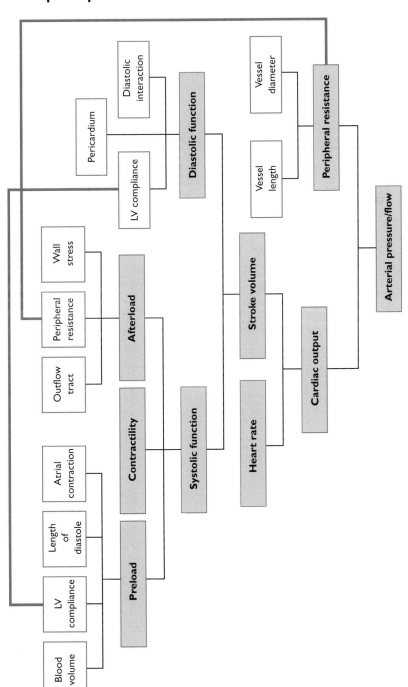

Fig. 8 Interactions between cardiovascular parameters
LV, left ventricle.

2

Cardiac

Cardiac output (CO, Q_s) = heart rate (HR) × stroke volume (SV)

SV = LV end-diastolic (LVED) volume − LV end-systolic volume

$$\text{Cardiac output} = \frac{\text{arterial pressure}}{\text{systemic vascular resistance (SVR)}}$$

$$\text{SVR} = \frac{\text{aortic pressure (AoP)} - \text{right atrial pressure (RAP)}}{Q_s}$$

$$\text{LV wall stress} = \frac{\text{LV systolic pressure} \times \text{LVED radius}}{\text{LV wall thickness}}$$

$$\text{Ejection fraction (EF)} = \frac{\text{EDV} - \text{ESV}}{\text{EDV}}$$

where EDV and ESV are end-diastolic and end-systolic volumes respectively.

Measurement of cardiac output

1 Fick principle:

$$Q_s \text{ or CO (L min}^{-1}) = \frac{O_2 \text{ uptake by lung (mL min}^{-1})}{\text{arterial–venous difference in } O_2 \text{ content}}$$

2 Dye dilution test. A known quantity of dye (indicator) is injected as a bolus into the circulation and a concentration–time curve is sampled downstream (usually in a peripheral artery).

$$\text{CO} = \frac{\text{amount of indicator injected (mg)}}{\text{mean indicator concentration (mg mL}^{-1})} \times \text{curve duration}$$

3 Thermodilution. Cooled saline (indicator) is injected into the right atrium. Blood temperature changes are monitored by a thermistor on the tip of a pulmonary artery catheter. A thermal dilution curve, analogous to the concentration–time curve above, is plotted. Because the body acts as a heat sink, there is no problem of indicator recirculation. Rapid successive measurements are possible.

4 Two dimensional echocardiography (approximate, non-invasive).

5 Nuclear medicine multiple uptake gated acquisition (MUGA) scan (very accurate, time consuming).

2

Blood flow

Flow through a vascular bed is determined by the pressure gradient across it and vascular resistance (Ohm's law).

$$\text{Flow} = \frac{P}{R}$$

(P, perfusion pressure; R, resistance)

Measurement
- Fick principle (invasive, very accurate).
- Plethysmography (limbs only).
- Doppler (non-invasive, individual vessels only): flow = $V \times SA$. (V, velocity; SA, surface area, from Doppler.)

Vascular resistance (Poiseuille–Hagen formula)

Vascular resistance is directly proportional to the viscosity of blood and length of blood vessel, and inversely proportional to the (vessel radius)^4 (Poiseuille's law).

$$R = \frac{8\eta L}{\pi r^4}$$

(η, blood viscosity; L, vessel length; πr^4, vessel surface area $\times$ radius^2)

Clinical importance
1 Small changes in arteriolar radius (sympathetic drive) lead to dramatic changes in systemic vascular resistance making it a highly potent system.
2 Changes in viscosity due to polycythaemia etc. significantly reduce organ blood flow (see above).

Wall tension (Laplace's law)

$$T = \frac{P \cdot r}{Th}$$

(P, perfusion pressure; r, radius; Th, wall thickness)

Clinical importance
1 Cardiac wall dilatation leads to ↓ wall tension (decreased function).
2 Cardiac wall hypertrophy leads to ↑ wall tension (increased function).
3 ↑ Aortic aneurysm radius leads to ↑ wall tension at a given blood pressure (increased likelihood of leak).

Cardiac cycle I

The cardiac cycle is best divided into seven phases as follows:

LV contraction

(b) *Isovolumetric contraction*: both aortic valve (AoV) and mitral valve (MV) are closed, LV volume is fixed. S1 occurs midphase.

(c) *Rapid ejection*: begins on AoV opening.

LV relaxation

(d) *Start of relaxation and reduced ejection*: LV to Ao flow maintained by aortic distensibility (windkessel effect).

(e) *Isovolumetric relaxation*: AoV and MV closed.

(f) *Rapid LV filling*: begins on MV opening, causes S3.

(g) *Slow LV filling*: as atrial and ventricular pressures equalize.

(a) *Atrial systole.*

- Physiological systole extends from the start of isovolumetric contraction (b) to the end of rapid ejection (c), i.e. peak of the ejection phase.
- Cardiological systole extends from S1 (MV closure) to S2 (AoV closure).

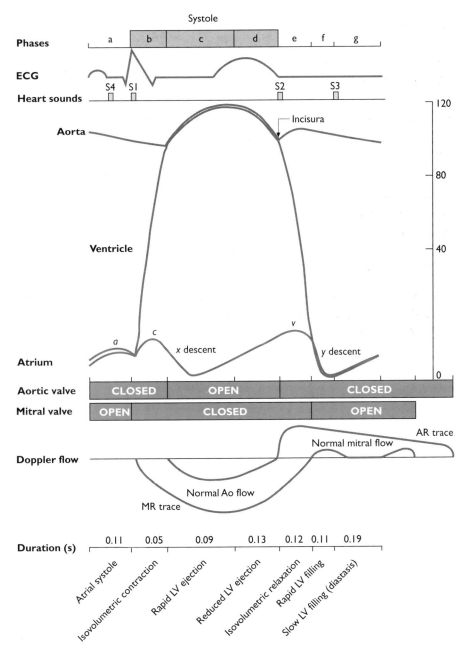

Fig. 9 Electromechanical relationships
AR, aortic regurgitation; ECG, electrocardiogram; MR, mitral regurgitation.

Cardiac cycle 2

Heart sounds

• S3 coincides with y descent (due to limitation of LV longitudinal expansion during early diastole).
• S4 occurs during ventricular filling following atrial contraction and coincides with the a wave.

Aortic pressure wave

• The incisura marks closure of the AoV (S2) and the end of systole.

Atrial pressure wave

• a wave coincides with atrial systole.
• c wave is a small positive deflection due to protrusion of the closed mitral (or tricuspid) valve into the atrium.
• x descent coincides with atrial relaxation and descent of the atrial floor during ventricular systole.
• v wave is due to continued atrial filling against a closed mitral (or tricuspid) valve.
• The v wave is higher than the a wave in the left atrium because of pulmonary vein contraction.
• y descent is related to the decline in atrial pressure when the closed mitral (or tricuspid) valve opens.

Pressure–volume loop

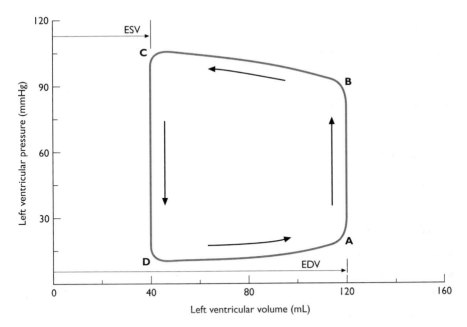

Fig. 10
The cardiac cycle proceeds in an anticlockwise direction. (A) End diastole; (B) AoV opening; (C) AoV closure; (D) MV opening. Segments AB, BC, CD and DA represent isovolumetric contraction, ventricular ejection, isovolumetric relaxation and diastolic filling respectively. EDV and ESV are represented by points A and C respectively. The area enveloped by the loop represents the stroke work.

    Segment DA (the end-diastolic pressure–volume relationship, passive tension curve) is altered in diastolic dysfunction.

Electrophysiology

Conduction system

Structure	Location (blood supply)	Autonomic innervation	Physiology
SA node	SVC RA junction (RCA 55%, LCX 45%)	Abundant	Pacemaker
AV node	Inter-atrial junction (RCA 90%, LCX 10%)	Abundant	Conduction delay Subsidiary pacemaker
His bundle	Membranous septum (LAD)	Sparse	Conduction from AV node to bundle branches
Bundle branches	Muscular septum, ventricles (LAD, RCA)	Sparse	Ventricular activation

AV, atrioventricular; LAD, left anterior descending; LCX, left circumflex artery; RA, right atrium; RCA, right coronary artery; SA, sinoatrial; SVC, superior vena cava.

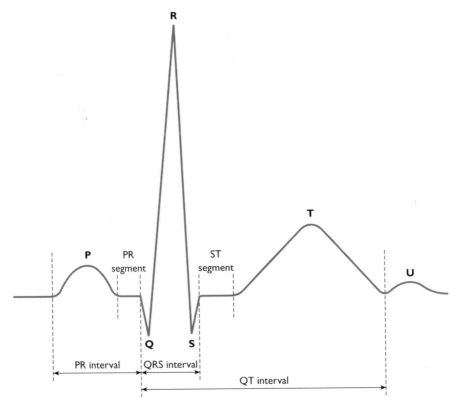

Fig. 11 Various electrocardiographic complexes and intervals (lead II)

Approach to 12-lead electrocardiogram

1 Rate	Estimated from RR interval
2 Mean frontal electrical axis	Assessed by measuring the amplitudes of QRS complexes in limb leads
	Normal axis 0–90° (complexes in I and II are predominantly positive)
	The axis lies at 90° to an isoelectric complex (where positive and negative deflections are equal)
	Axis in the horizontal plane can be determined using the precordial leads. The normal transitional complex should lie between V2 and V4. Clockwise shift of the transitional complex towards V5 and V6 suggests a posterior axis. Anticlockwise shift of the transitional complex towards V1 and V2 suggests an anterior axis
3 P wave	Atrial systole
	Best seen in leads II and V1
	Normal height 0.25 mV (2.5 mm)
	Normal width 0.12 s (3 mm)
4 PR interval	Normal AV conduction time 0.12–0.2 s (3–5 mm)
	> 0.2 s implies conduction block
	< 0.12 s implies an abnormal atrial focus or accessory pathway
5 Rhythm	Sinus rhythm is implied by:
	1 P wave of sinus origin
	2 constant normal PR interval
	3 constant P wave morphology in a lead
	4 constant RR interval
	Respiratory sinus arrhythmia (increase in rate with inspiration) is common in children and young adults (see p. 178)
6 QRS complex	Normal ventricular depolarization time < 0.12 s (3 mm)
	Septal activation from L to R produces small physiological Q waves in V4 to V6
	Q waves must be at least 0.04 s in duration and 25% of the height of the following R wave to be pathological
	LV wall depolarization produces S and R waves in right and left precordial leads respectively
	Finally RV wall depolarization produces opposite deflections
7 QT interval	Ventricular depolarization and repolarization time ranges from 0.35 to 0.43 s
	Corrected according to rate
	$QT_c = QT/\sqrt{(RR)}$
	QT interval differs from lead to lead (QT dispersion)
8 ST segment	Extends from J point (end of S wave) to beginning of T wave
	Point of deviation is 0.06 s (1.5 mm) from J point during exercise
9 T wave	T wave axis should not vary from QRS axis by > 45°
	Inverted T waves in I, II, V4 to 6 are abnormal
	Flattened and followed by U wave in hypokalaemia

Shunt physiology

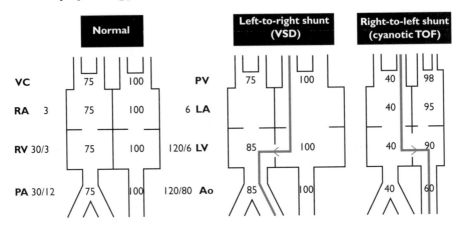

Fig. 12 Shunt physiology
Numbers within cardiac chambers represent oxygen saturations. Ao, aorta; LA, left atrium; PA, pulmonary artery; PV, pulmonary vein; RV, right ventricle; TOF, tetralogy of Fallot; VC, vena cava; VSD, ventricular septal defect. (Adapted from Park, M.K. (1984) *Pediatric Cardiology for Practitioners*. Year Book Medical Publishers, Chicago.)

Left-to-right shunt	Right-to-left shunt
Increases pulmonary blood flow compared to systemic blood flow	Increases systemic blood flow compared to pulmonary blood flow
$Q_p/Q_s > 1$	$Q_p/Q_s < 1$
S_aO_2 normal	S_aO_2 falls

The Fick principle used to determine CO is also applicable to shunt quantification.

$$Q_p \text{ (pulmonary blood flow, L min}^{-1}) = \frac{O_2 \text{ uptake (mL min}^{-1})}{P_V O_2 - P_a O_2}$$

$$Q_s \text{ (systemic blood flow, L min}^{-1}) = \frac{O_2 \text{ delivery (mL min}^{-1})}{S_a O_2 - M_V O_2}$$

where $P_V O_2$, $P_a O_2$, $S_a O_2$ and $M_V O_2$ refer to O_2 content (in millilitres of O_2 per litre of blood) in pulmonary venous, pulmonary arterial, systemic arterial and mixed venous blood.

Thus,

$$\frac{Q_p}{Q_s} = \frac{S_a O_2 \% - M_V O_2 \%}{P_V O_2 \% - P_a O_2 \%}$$

arterial-mixed venous saturation difference is approximately 25% while $P_V O_2 \%$ is about 98% in normal subjects.

$$PVR = \frac{PAP - PVP}{Q_p} \text{ (Ohm's law as applied to the circulation)}$$

where PVR is pulmonary vascular resistance, PAP is pulmonary arterial pressure and PVP is pulmonary venous pressure.

Similarly,

$$SVR = \frac{AoP - RAP}{Q_s}$$

where SVR is systemic vascular resistance, AoP is aortic pressure and RAP is right atrial pressure.

Thus,

$$\frac{PVR}{SVR} = \frac{Q_s}{Q_p} \times \frac{PAP - PVP}{AoP - RAP}$$

Congenital heart disease physiology

A physiological approach to the patient based on colour.

Pink	Pink or blue	
(a) Left-to-right shunts	**(b) Pulmonary venous (PV) congestion**	**(c) Prejudiced systemic perfusion**
These result in the volume overload of chambers receiving excess blood, and ultimately failure of the corresponding ventricle. The chest radiograph typically shows pulmonary plethora. Examples include atrial septal defect (ASD), VSD and patent ductus arteriosus (PDA)	Obstruction of PV return can occur at mitral/submitral level (LV inflow obstruction), within the atrium (cor triatrium) or within the PV pathway (obstructed total anomalous pulmonary venous drainage (TAPVD))	Due to low stroke volume (as in hypoplastic left heart), left ventricular outflow tract obstruction (LVOTO) (aortic stenosis (AS)) or coarctation of the aorta

Blue		
(d) Low pulmonary blood flow	**(e) Transposition streaming**	**(f) Intracardiac mixing**
Patients are not breathless, lung fields appear oligaemic on radiography. Examples include right ventricular outflow tract obstruction (RVOTO) and TOF.	Systemic and pulmonary circuits occur in parallel rather than in series. Survival depends on the extent of mixing. Breathlessness suggests an additional lesion such as a VSD	Patients are usually breathless. Mixing can occur at atrial level (unobstructed TAPVD), at ventricular level (all univentricular hearts) or in great arteries (truncus arteriosus)

Tetralogy of Fallot

Degree of cyanosis depends on relative resistance to RV and LV outflow. For example, cyanosis is worse on exertion when SVR falls due to peripheral vasodilatation.

2

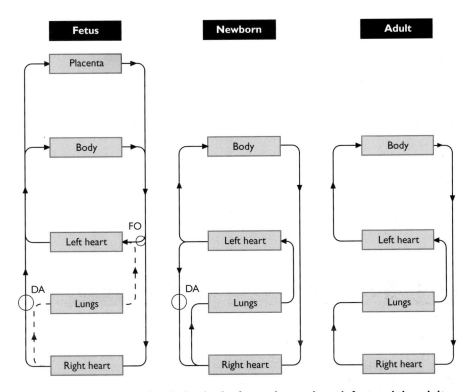

Fig. 13 Diagram of the circulation in the fetus, the newborn infant and the adult
DA, ductus arteriosus; FO, foramen ovale. (Redrawn from Born, G.V.R. *et al.* (1954) Changes in the heart and lungs at birth. *Cold Spring Harbor Symposia on Quantitative Biology* **19**, 102.)

Control of peripheral circulation

2

Vasodilatation (vasorelaxation)

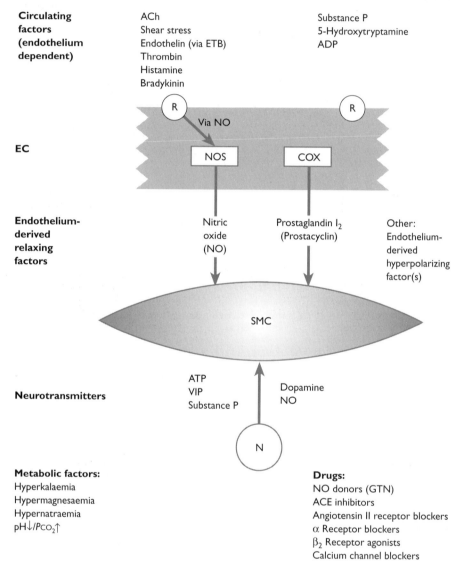

Circulating factors (endothelium dependent)	ACh Shear stress Endothelin (via ETB) Thrombin Histamine Bradykinin	Substance P 5-Hydroxytryptamine ADP

EC

| **Endothelium-derived relaxing factors** | Nitric oxide (NO) | Prostaglandin I_2 (Prostacyclin) | Other: Endothelium-derived hyperpolarizing factor(s) |

SMC

| **Neurotransmitters** | ATP
VIP
Substance P | Dopamine
NO |

N

Metabolic factors:
Hyperkalaemia
Hypermagnesaemia
Hypernatraemia
$pH\downarrow/P_{CO_2}\uparrow$

Drugs:
NO donors (GTN)
ACE inhibitors
Angiotensin II receptor blockers
α Receptor blockers
β_2 Receptor agonists
Calcium channel blockers

Fig. 14
ACE, angiotensin-converting enzyme; ACh, acetylcholine; COX, cyclooxygenase; EC, endothelial cell; ETB, endothelin type B receptor; GTN, glyceryl trinitrate; N, perivascular nerve; NOS, nitric oxide synthetase; SMC, smooth muscle cell; VIP, vasoactive intestinal peptide. R, specific cell surface receptor.

Vasoconstriction

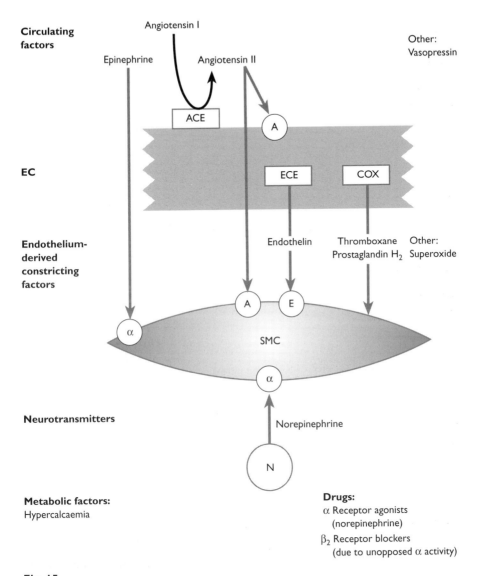

Fig. 15
ECE, endothelin-converting enzyme. Receptors: A, angiotensin; E, endothelin.

Renin–angiotensin system (RAS)

Actions of angiotensin II (AII) and tissue distribution of angiotensin receptors

	A I R	AIIR
Heart	Increased contractility, ventricular hypertrophy	Inhibition of collagenase
Vasculature	Vasoconstriction, vascular hypertrophy, angiogenesis	Inhibition of angiogenesis
Kidneys	Reduced GFR due to vasoconstriction (efferent > afferent), mesangial cell contraction and collagen synthesis, inhibition of renin release, increased proximal Na transport	
Brain	PGE_2 and vasopressin release	
Autonomic nervous system	β Adrenergic stimulation, vagal suppression	
Adrenal glands	Aldosterone biosynthesis, catecholamine secretion	

A I R, angiotensin I receptor; AIIR, angiotensin II receptor; GFR, glomerular filtration rate; PGE_2, prostaglandin E_2.

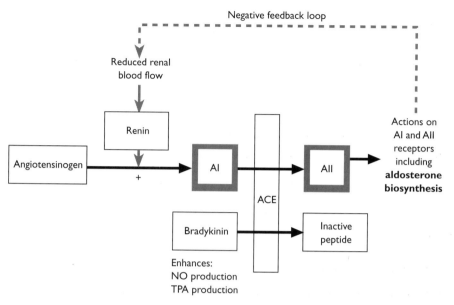

Fig. 16 Renin–angiotensin system
TPA, tissue plasminogen activator.

ACE DD (deletion) polymorphisms are associated with an increased risk of myocardial infarction (MI) as compared to II (insertion) or insertion/deletion genotypes.

The RAS has been implicated in the pathogenesis of hypertension. Hypertension due to renal artery stenosis accounts for 1% of hypertensives. The juxtaglomerular apparatus of the affected kidney responds to decreased perfusion pressure with increased renin secretion.

ACE inhibitors (contraindicated in renal artery stenosis) increase plasma renin further by their hypotensive effect and by interrupting the feedback inhibition of renin release by AII. ACE inhibitors result in a fall in AII, a rise in bradykinin production and a rise in NO and prostacyclin release.

Angiotensinogen variants are associated with hypertension. Oral contraceptive-induced hypertension may be due to increased secretion of angiotensinogen.

Nitric oxide

Nitric oxide (NO) is:
- produced continuously by the action of NOS on L-arginine by endothelial and other cells;
- ADMA (asymmetrical dimethylarginine) is an endogenous inhibitor of NOS;
- a free radical by virtue of its unpaired electron and therefore highly reactive.
  There are three isoforms of NOS.

	Endothelial NOS (eNOS)	Macrophage inducible NOS (iNOS)	Neuronal NOS (nNOS)
Chromosome	7	17	12
Activated by	Fluid flow, endothelial shear stress, ACh, serotonin, thrombin, ADP and substance P		Glutamate binding to postsynaptic terminals in hippocampal pyramidal cells
Induced by	Physical exercise	IL-1; IFNγ; TNFα; OxLDL	
Inhibited by		IL-4, 8 and 10; TGFβ; IGF-1; PDGF	
Amount released	Small, pulses	Large, continuous	
Comments	Also produced by kidney tubular epithelial cells		

IFNγ, interferon-γ; IFG-1, insulin-like growth factor I; IL-1, interleukin 1; OxLDL, oxidized low-density lipoprotein; PDGF, platelet-derived growth factor; TGFβ, transforming growth factor-β; TNFα, tumour necrosis factor-α.

Actions of nitric oxide

- *Vasodilatation.* NO diffuses across the EC membrane into SMCs, activating guanylate cyclase (GC), which in turn converts GTP to cGMP. This leads to SMC relaxation. NO production in response to shear stress may play a role in autoregulation.
- *Other vascular effects.* Regulates endothelial cell migration and proliferation. Regulates vascular remodelling (see p. 39).

- *Haemostasis.* Prevents platelet aggregation and adhesion. This negative feedback mechanism localizes clot formation and vasoconstriction to sites of endothelial damage.
- *Host defence.* Prevents leukocyte and monocyte adhesion. NO combines with O_2 to generate toxic reactive oxygen species.
- *Neurological actions.* Neurotransmitter in nonadrenergic noncholinergic nerves. Possible role in synaptic plasticity and memory (long-term potentiation in the hippocampus and long-term depression in cerebellum). Responsible for the relaxation of corpus cavernosum and penile erection, and anal sphincter relaxation (NO released onto adventitial side of blood vessels). Mechanism underlying reperfusion injury in stroke may involve excessive glutamate and therefore NO production.

Regional variation in nitric oxide synthesis

- Resistance vessels: big basal NO synthesis. NO synthesis is particularly high in those vessels (100–200 μm diameter) best placed to control distribution of flow and where shear forces are maximal.
- Veins: no basal NO synthesis.
- Conduit vessels: small basal NO synthesis.

   The major site of nitrovasodilator drug action is on the venous system. This reflects more efficient metabolism of the drug to its active component NO rather than endogenous NO production.

Endothelins

- Potent vasoactive peptides secreted by the basal surface of the EC towards the underlying SMC.
- Act locally in an autocrine and paracrine fashion.

Types

1 Endothelin (ET) 1 (equipotent for endothelin type A (ETA) and endothelin type B (ETB) receptors).
2 ET2.
3 ET3 (predominantly acts on ETB receptors).

Receptors (members of rhodopsin superfamily)

ETA	ETB
ET1 = ET2 > ET3	ET1 = ET2 = ET3
High affinity	Low affinity
Expressed on SMC, myocardial cells, mesangial cells	Expressed on EC, glomerular cells and vasa recta

Actions

Cardiovascular	Renal
Initial vasodilatation mediated by NO release followed by sustained vasoconstriction	Increased renal vascular resistance
Vasoconstriction of coronary (not pulmonary) arteries	Reduced GFR (afferent and efferent arteriolar constriction)
Vascular SMC proliferation	Inhibition of Na/K-ATPase
Positive inotropic effect on cardiac atria	Mesangial cell proliferation
Inhibition of renin release and stimulation of ACE	Increased mesangial matrix synthesis
Modulation of blood volume by release of ANP and aldosterone	

ANP, atrial natriuretic peptide.

Pharmacology

Cyclosporin upregulates endothelin receptors and may account for cardiotoxicity.

Blood pressure

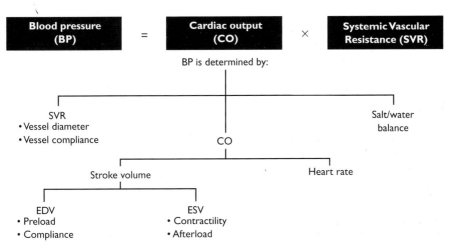

Fig. 17

Control of blood pressure

Short to intermediate term

1 Local
- Myogenic autoregulation (contractile response to stretch. As vessel diameter increases, transmural pressure falls to keep wall tension constant as per Laplace's law;
- Stress relaxation (due to a delayed increase in vessel compliance);
- Autocrine and paracrine mechanisms (typically endothelium dependent):
  vasorelaxation (due to NO, prostacyclin, see p. 30)
  vasoconstriction (due to endothelin, RAS, see p. 31)

2 Baroreceptor reflexes

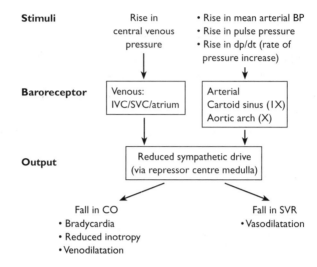

Fig. 18a

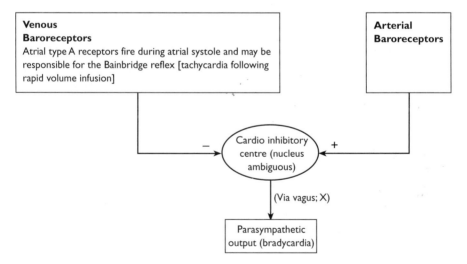

Fig. 18b

Other cardiac receptors

1 Atrial type B receptors discharge during late in atrial diastole or relaxation and sense changes in venous return.

2 Ventricular receptors respond to stretch during isovolumetric ventricular contraction and may maintain vagal showing of heart rate at rest.

Long term

Vascular remodelling
- Endothelium and NO play a central role.
- Active process of structural alteration resulting from a balance between cell growth, cell death, migration, matrix formation and degradation.
- Remodelling is dependent on haemodynamic stimulation, local growth factor production and vasoactive substances.
- For example, in response to increased arterial pressure, the ratio of arterial wall thickness to luminal diameter is increased.

Renal regulation of fluid and Na balance (see p. 84)

Physiological approach to mechanisms of hypertension

Excess cardiac output
- Beta adrenergic drive
  e.g. emotional stress with tachycardia

Excess peripheral resistance
- Endothelial dysfunction
  (increased endothelin and decreased NO production)
- Alpha adrenergic drive
- Renin angiotensin system activation
  Abnormal vascular remodelling

Abnormal salt/water balance
- Renal failure
- Abnormal steroid biosyntheses
  (e.g. aldosterone/glucorticoid excess)
- Abnormal steroid biosynthesis
  (e.g. aldosterone/glucocortoid excess)
- Abnormal Na transport
  (e.g. ENAC mutation in Liddle's syndrome)

NB: Because $BP = CO \times SVR$, sustained hypertension due to an abnormality in either entity results when baroreceptor control is inadequate.

Responses to hypovolaemia

Each mechanism of compensation has a different relative efficacy and range of blood pressure over which the mechanism is operative. These parameters determine how important each mechanism is at different times during hypovolaemia.

Speed of onset	Mechanism	Relative maximum effect	Range of pressures active over (mmHg)
Acute	Baroreceptor-mediated veno- and vasoconstriction	0.7	50–170
	CNS ischaemic response	1.2	0–50
Subacute	ADH production/ANP reduction	0.3	50–150
	Myogenic response	0.25	0–200
	RAS vasoconstriction	0.2	50–100
	Aldosterone production	0.4	40–150
	Capillary fluid shifts	0.2	0–200
Chronic	Albumin fluid shifts	0.4	–
	Overall renal fluid retention control	1.6	–

ADH, antidiuretic hormone; CNS, central nervous system.

Notes

1 Vasomotor centre (VMC):
   - located in the dorsal medullary reticular formation;
   - depressor area responds to baroreceptor stimulation;
   - pressor area responds to fear, pain, hypoxia, chemoreceptor stimulation;
   - net outflow modulates sympathetic vasomotor tone.
2 CNS ischaemic response:
   - most potent early response to hypotension;
   - only active over a very low range of pressures;
   - effectively a 'last ditch' attempt at maintaining CNS blood flow.
3 Renal fluid control:
   - the most effective.

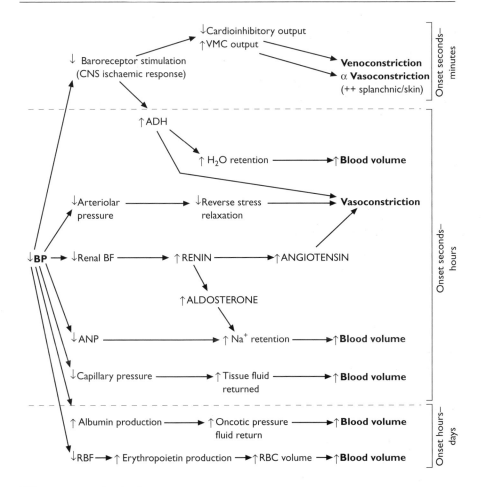

Other responses involved

↑Glucocorticoid response facilitates the activity of:
  • catecholamines
  • ADH
  • aldosterone
  • erythropoietin

↑Marrow hypoxia produces increased development and release of:
  • platelets
  • clotting factors

Fig. 19 Physiological responses to hypovolaemia
BF, blood flow; RBC, red blood cell; RBF, renal blood flow.

Responses to exercise

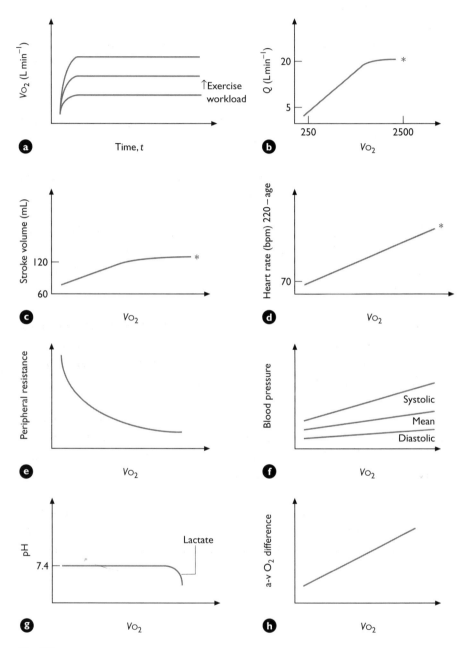

Fig. 20
a-v, arteriovenous; *, maximal exercise.

(a) Vo_2. The rate of consumption of O_2 rises rapidly with the onset of exercise with a level reached in steady exercise which is proportional to the overall workload. Used as the indicator of 'level of exercise'.

(b) Q *(cardiac output L min⁻¹)*. The overall limiting factor of the maximum level of steady aerobic exercise possible (maximum rate of delivery of oxygenated blood to tissues).

(c) *Stroke volume*. Rises in response both to increased venous return and to rising sympathetic drive (inotropism).

(d) *Heart rate*. Rises mainly in response to increased cortically modulated sympathetic activity (chronotropism).

(e) *Peripheral resistance*. Falls rapidly due to increasing vasodilatation of muscle vessels in response to local factors which increase as exercise workload increases. Splanchnic and renal vasoconstriction via sympathetic tone prevent an even more dramatic fall in peripheral resistance.

(f) *Blood pressure*. Systolic pressure rises due to the fact that it principally reflects stroke volume which is rising markedly. Diastolic pressure also rises or is held constant despite marked falls in peripheral resistance since this is outweighed by rising stroke volume and heart rate.

(g) *pH*. During normal exercise, pH remains constant due to aerobic metabolism with increased production of CO_2 being matched by increased respiratory removal. Only when anaerobic exercise overloads the liver's ability to handle the lactate produced does pH fall.

(h) *Arteriovenous oxygen difference*. Represents increased O_2 extraction from the blood (particularly in exercising muscles) due to lowered intracellular Po_2 and thus steeper O_2 gradient into the cell from the capillary.

Effects of conditioning

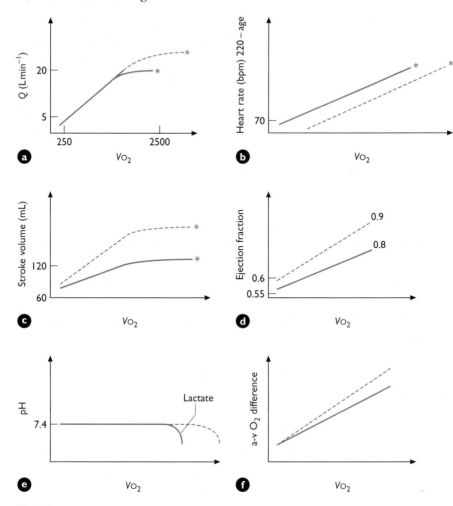

Fig. 21
*, Maximal exercise.

(a) *Cardiac output.* This increases as a result of increased stroke volume at a given level of exercise and an increase in the maximum heart rate attainable. This also results in a more O_2-efficient cardiac output. This is because at a given cardiac output, the O_2 requirement of the myocardium is lowered, since the increase in O_2 used by the increased stroke volume is more than off-set by the fewer beats required to produce that level of output.

(b) *Heart rate.* Maximal heart rate increases due to improved maximal sympathetic response. In addition, since stroke volume increases even at low levels of exercise, this allows the required cardiac output to be achieved at lower heart rates than in unconditioned individuals.

(c and d) *Stroke volume/ejection fraction.* The most important element of training and conditioning for the heart is the ventricular wall hypertrophy which is undergone and the increased wall lusiotropy (pliability). This allows the maximal end-diastolic volume to increase both at rest and during exercise. This improves myocardial performance simply by virtue of Starling's law. In addition, the improved contractility increases the stroke volume by increasing the ejection fraction as exercise progresses, improving the efficiency of each 'pump'.

(e) *pH.* The level of work at which pH begins to fall rises due to improved muscle blood flow by virtue of increased cardiac output.

(f) *Arteriovenous oxygen difference.* Improved muscle tissue oxygen flow due to further lowering of intracellular P_{O_2} during exercise occurs due to increased myoglobin levels which maintain a lowered free P_{O_2} whilst providing a reservoir of available O_2 at very low O_2 pressures.

Cardiac dysfunction

Myocardial infarction

Damage to a ventricle or an increase in the resistance against which it ejects (afterload) shifts the preload–stroke volume relationship downward and to the right. Inotropic stimulation or afterload reduction shifts the equation upward and to the left. Hence, in anterior or lateral infarction (where the LV is disproportionately damaged) or increased SVR, RV and LV relationships separate. This increases the risk of pulmonary oedema at low RA pressures. Worsening RV function (as in inferior infarction), on the other hand, brings the relationships closer together. Pulmonary oedema does not occur at high RA pressures.

Cardiac output can be improved by:
1 atrial pacing (to increase the heart rate);
2 optimizing the filling pressures (whilst avoiding pulmonary oedema);
3 improving contractility with inotropes;
4 afterload reduction (with intravenous nitrates and/or counterpulsation with an intra-aortic balloon pump).

With extreme impairment of ventricular function, the preload–stroke volume relationship becomes flat. In these cases, raising the filling pressures on both sides of the heart has no effect on cardiac output and results in pulmonary oedema.

Cardiac tamponade

Fluid accumulation in the pericardial space results in a rise in pericardial pressure throughout the cardiac cycle and a rise in atrial pressures to maintain a normal transmural pressure across ventricular walls at end diastole. Compensatory mechanisms include tachycardia, increasing SVR to maintain BP (although CO falls) and reduced ANP secretion. Sinus node (SN) ischaemia produces a bradycardia in late tamponade.

Pulsus paradoxus (reduction in arterial and pulse pressure with inspiration) is an exaggeration of normal physiology (a fall of 10 mmHg is thought to be significant in the absence of bronchospasm). This is thought to be because pulmonary venous pressure fails to rise appropriately in inspiration, falls below that in the pericardium and results in a fall in LV filling. Furthermore, the normal inspiratory increase in systemic venous return means that RV volume can only be maintained at the expense of septal deviation and LV cavity obliteration. There is an abnormal inspiratory rise in tricuspid valve (TV) flow and a > 15% reduction in MV flow.

Chronic cardiac failure

Responses to impaired cardiac performance

Response	Short-term effects	Long-term effects
	(Mainly adaptive) (Haemorrhage, acute heart failure)	(Mainly deleterious) (Chronic heart failure)
1 Salt and water retention	Augments preload	Pulmonary congestion
2 Vasoconstriction and vascular remodelling	Maintains cerebral and coronary perfusion	Increases afterload Increases cardiac energy expenditure
3 Sympathetic activation	Increases heart rate and CO	Increases energy expenditure Downregulation of β adrenoceptor density and function and reduced catecholamine sensitivity
4 Hypertrophy and ventricular remodelling	Unloads individual muscle fibres	Cardiac cell death
5 Prolonged action potential		Increases contractility and energy expenditure

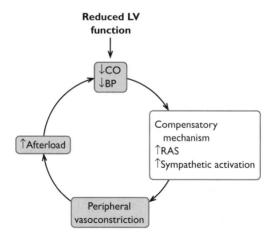

Fig. 22

Valsalva manoeuvre

The manoeuvre during which a deep breath is inhaled and the intrathoracic pressure raised by raising the intra-abdominal pressure whilst keeping the glottis closed (e.g. during defaecation, difficult micturition, parturition and heavy lifting).

Venous return

Venous return is reduced by compression of the intra-abdominal and intrathoracic vena cavae with the venous blood being retained within the capacitance vessels of, particularly, the abdomen.

Blood pressure

After an initial rise due to the effect of a sudden tensing of muscles and compression of abdominal arteries during the onset of straining, reduced venous return reduces right, and to a lesser extent left, ventricular diastolic pressure with a resultant fall in stroke volume. This reduces particularly systolic pressure. Diastolic pressure is maintained by a progressive rise in peripheral resistance initiated by lowered baroreceptor stimulation. Systolic pressure is partially maintained by a progressive tachycardia.

After release of straining, the sudden release of the accumulated venous blood causes a marked increase in the end-diastolic pressure and volume with an increase in stroke volume. The tachycardia and vasoconstriction present cause this increased cardiac output to result in a sudden increase of arterial pressure above normal with a widened pulse pressure due to the underlying tachycardia.

Heart rate

Rise mediated by baroreceptor stimulation with falling cardiac output as venous return drops. Return to normal is more rapid than the fall in vasoconstriction thus returning pulse pressure to normal quite rapidly after release of straining.

Peripheral resistance

Rises progressively due to falling baroreceptor stimulation but returns to normal more slowly than other parameters after release of straining to cause a moderately prolonged 'post-strain' diastolic hypertension.

2

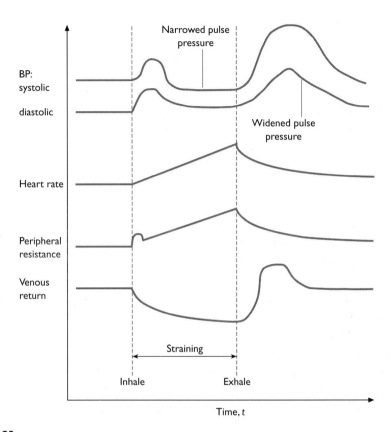

Fig. 23

3: RESPIRATORY PHYSIOLOGY

3

Basic principles

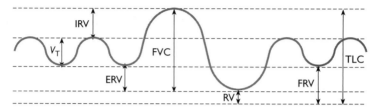

Fig. 24 Spirometry
FRV, functional residual volume; for definition of other abbreviations, see table below.

	Typical adult values		
	Males (mL)	Females (mL)	Definition
Tidal volume (V_T)	500	500	Amount of air entering (leaving) lungs with normal inspiration (expiration)
Inspiratory reserve volume (IRV)	3300	1900	Volume of air inspired with maximal inspiratory effort in excess of V_T
Expiratory reserve volume (ERV)	1000	700	Volume of air expired with maximal expiratory effort in excess of V_T
Residual volume (RV)	1200	1100	Volume of air left in lung after maximal expiratory effort
Forced vital capacity (FVC)	4800	3100	Volume of air expelled by maximal expiration from full maximal inspiration = IRV + V_T + ERV
Total lung capacity (TLC)	6000	4200	TLC = FVC + RV

Measurement of respiratory volumes

TLC

• Helium dilution technique.

Dead space

• Anatomical: fast response CO_2 measurements (Fowler's method).
• Physiological: Bohr equation derived

$$V_{dead} = V_T - \frac{P_E CO_2}{P_A CO_2} \cdot V_T$$

($P_E CO_2$, mean end-tidal CO_2 partial pressure; $P_A CO_2$, alveolar CO_2 partial pressure)

Lung parameters

- Shunt equation

$$\frac{Q_{sh}}{Q_t} = \frac{P_A O_2 - P_a O_2}{P_A O_2 - P_v O_2}$$

- Dead space equation—see above

Diffusing capacity

CO gas transfer for the total lung (DLCO) is expressed as the amount of CO transferred per minute per unit partial pressure gradient of CO (mmol min^{-1} kPa^{-1}).

DLCO across the alveolar capillary membrane is measured by the single breath CO method. A full breath of gas mixture containing 0.3% CO is inhaled and the amount of gas transferred after breath-holding for 10 s at TLC is measured. CO has a 200-fold higher affinity for Hb than O_2, so its dilution in the exhaled gas mixture reflects the diffusing capacity of the lung.

DLCO is corrected for alveolar volume (VA) to give the transfer coefficient (KCO).

Causes of changes in KCO

Low (low DLCO, normal/ raised VA)	Normal (low DLCO, low VA)	High (raised DLCO)
• Emphysema • Pulmonary vascular disease • Anaemia	• Interstitial lung disease • Pneumonectomy	• Increased pulmonary blood flow (left-to-right shunt, exercise, altitude, supine posture) • Alveolar haemorrhage • Polycythaemia • Bronchial asthma • Chest wall abnormality normal DLCO, low VA

Gas transport

Normal values/definitions

The amount of oxygen in blood can be represented as:
- O_2 = (really haemoglobin (Hb) saturation) S_aO_2 = [HbO_2]/([Hb] + [HbO_2]).
- O_2 content = (O_2 saturation × Hb × carrying capacity (1.39)) + dissolved O_2 (same for all sizes and ages; 20 mL O_2 dL^{-1}).
- O_2 capacity (mL L^{-1}) = maximum O_2 Hb able to carry (= 1.34 mL g^{-1} of Hb × [Hb] g L^{-1}).
- O_2 actually carried = O_2 saturation (%) × O_2 capacity {N sat = 97%}.
- O_2 dissolved in plasma (mL L^{-1}) = P_aO_2 × 0.03 {N = 2.9 mL L^{-1}}.

Partial pressures

	O_2 (kPa)	O_2 (mmHg)	CO_2 (kPa)	CO_2 (mmHg)
Inspiratory partial pressure, P	20	150	27	0.2
Alveolar partial pressure, P_A	13.3	100	5.3	40
Arterial partial pressure, P_a	12.6	95	5.3	40

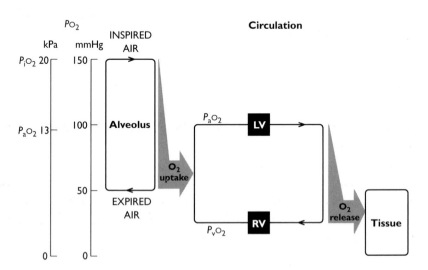

Fig. 25 Model for oxygen transport
LV, left ventricle; RV, right ventricle.

Alveolar–arteriolar oxygen difference ($P_{(A-a)}O_2$)

- P_AO_2 (alveolar O_2 tension) is estimated using the alveolar gas equation: $P_AO_2 = P_IO_2 - (P_aCO_2/R)$ or $P_AO_2 = P_IO_2 - (1.2 \times P_aCO_2)$.
- P_IO_2 at sea level is 20 kPa (150 mmHg)
- Upper limit for $P_{(A-a)}O_2$ is 2.5 kPa (18.5 mmHg) in normal young subjects and 4.7 kPa (35 mmHg) in subjects over 60 years of age (due to physiological shunting in bronchial and coronary circulations).
- $P_{(A-a)}O_2$ is usually normal with hypoxaemia due to hypoventilation. Hypercapnia is present in hypoventilation (suggested if $P_aO_2 + P_aCO_2 > 125$ mmHg (16.7 kPa)).

Respiratory quotient/exchange ratio (R)

R = volume CO_2 produced/volume O_2 consumed:
- 1.0 when only glucose is metabolized;
- normally 0.8;
- < 0.7 after 2-week starvation because of increased fat metabolism;
- the respiratory exchange ratio refers to whole-body exchanges while the respiratory quotient applies to the single cell.

Systemic oxygen delivery

Systemic oxygen delivery (DO_2) is the amount of oxygen delivered to peripheral tissues each minute.

DO_2 = arterial oxygen content × cardiac output

Systemic oxygen consumption (VO_2) is controlled by tissue metabolism and:
- increased during exercise, infection, and catecholamine and thyroid hormone release;
- decreased during rest, paralysis and hypothermia.

DO_2 is normally four to five times VO_2. A fall in DO_2 is compensated by increased extraction (manifest as a fall in venous saturation). When $DO_2/VO_2 < 2:1$, VO_2 decreases due to reaching the limit of increased extraction.

Oxygen carriage

- 98% carried bound to Hb.
- 2% carried in solution in the plasma.
- Hb:
  (a) four adjacent subunits of globin + haem bound to globin chain by salt bridge bond to histidine residue;
  (b) O_2 binding is via Fe^{2+} (ferrous iron) in haem moiety.
- At high pressures, flattening of the curve ensures that a fall in P_aO_2 due, for example, to lung disease does not result in profound desaturation.
- In the steep (middle) part of the curve, a small fall in P_aO_2, such as across the capillary bed, releases a greater amount of O_2 to tissues.
- P50 is the O_2 tension (mmHg) when binding sites are 50% saturated.

O_2 uptake
= (alveolar O_2 tension − P50) × diffusing capacity of lungs
= (P50 − mitochondrial O_2 tension) × diffusing capacity of tissue

Thus, if diffusing capacity of the lungs is low, P50 must fall to raise the pressure gradient in the lungs relative to that in the tissues and vice versa.

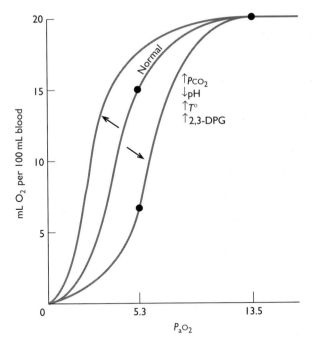

Fig. 26 Oxygen dissociation curve
2,3-DPG, 2,3-diphosphoglycerate.

Factors affecting the HbO₂ curve

Leftward shift	Rightward shift
Greater HbO₂ affinity (fall in P50)	Reduced HbO₂ affinity (rise in P50)
Enhances pulmonary O₂ loading	Enhances peripheral O₂ unloading
Caused by: • rise in pH • fall in temperature • fall in P_{CO_2} • fall in 2,3-DPG • fall in intracellular ATP • CO poisoning	Caused by: • fall in pH (acid Bohr effect) • rise in P_{CO_2} (CO_2 Bohr effect; hydrogen ions bind to Hb, stabilize deoxyHb and facilitate O₂ release to acidic tissues) • rise in 2,3-DPG • rise in intracellular ATP

Anaemia

- Primarily affects O_2 carriage by reduction of available Hb (hence reduction in O_2 capacity).
- No alteration of O_2 dissolved.
- No alteration of O_2 saturation at normal arterial P_aO_2.

Effects

- Reduced O_2 content at normal P_aO_2. (Leads to reduced O_2 pressure gradient across tissues and thus lowered driving gradient to O_2 delivery.)
- Lowered amount of O_2 released from binding for the same drop in venous PO_2.
- Markedly lowered P_vO_2 required to liberate the same amount of O_2.
- This causes more significant tissue hypoxia required to maintain the O_2 delivery gradient.
- Clinical effect negligible at Hb levels above 10 g dL^{-1}.
- Rapidly worsening clinical effects at Hb levels below 7 g dL^{-1}.

Clinical symptoms

- Faintness (central nervous system, CNS).
- Breathlessness.
- Lethargy.

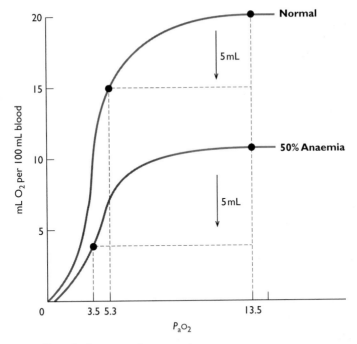

Fig. 27 Oxygen dissociation curve in anaemia

Carbon monoxide poisoning

Actions

- Binds to HbO_2-binding sites.
- Occupies binding sites with 230 times the affinity of O_2.
- Distorts unoccupied binding sites making O_2 binding to unoccupied sites less ready.
- Distortion of neighbouring sites occupied by O_2 causing *increased* haem–O_2 affinity and less ready tissue delivery.

Effects

- Lowered O_2 capacity.
- Less O_2 released to tissues.

These effects combined make 50% CO poisoning significantly more serious than 50% anaemia since, at normal tissue levels of hypoxia, very little O_2 is released and tissue hypoxia is much more intense.

Clinical symptoms

- Headache.
- Nausea.
- Lethargy.

Symptomatic at carboxyhaemoglobin (COHb) levels 30–50%.
Significant risk of tissue hypoxic injury at 50–70%.
Death at 70–75%.

Sources

- Internal combustion engine fumes.
- Incomplete coal burning.
- Burning of fossil fuels in O_2-restricted environments or with poor exhaust outlets.

Treatment

1 Attempted displacement of CO by 100% O_2 (maximal P_iO_2 = atmospheric pressure).
2 Accelerated displacement is effected by further raising P_iO_2 by administering hyperbaric 100% O_2 for relatively short periods of time.
3 Concomitant transfusion of warmed packed red cells or whole blood helps to restore available circulating O_2-binding sites.

3

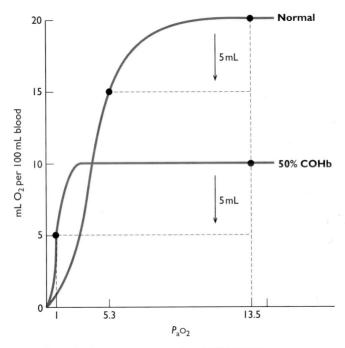

Fig. 28 Oxygen dissociation curve: normal and 50% COHb
Delivery of 5 mL of O_2 per 100 mL of blood from normal blood requires a fall to approximately 5 kPa Po_2. For 50% CO poisoning, this requires P_aO_2 to fall to approximately 1 kPa (usually incompatible with life).

The effects of altitude

Effects of ascent to altitude

Altitude (m)	Effects
0–2500	$\downarrow P_iO_2 \Rightarrow \downarrow P_aO_2$ Gradual hypoxia causing few symptoms and hypoxia has little effect on respiratory drive compared with the hypocapnia that any increase in ventilation causes, so hypoxia remains largely uncorrected
2500–4500	$\downarrow\downarrow P_iO_2 \Rightarrow \downarrow\downarrow P_aO_2$ Increasing effect of hypoxia overrides the hypocapnia of hyperventilation and respiratory rate (RR) rises progressively. Effect is mainly breathlessness
4500–6100	$\downarrow\downarrow\downarrow P_iO_2 \Rightarrow \downarrow\downarrow\downarrow P_aO_2$ Increase in ventilation becomes successively less effective at maintaining arterial oxygenation and more profound hypoxia ensues
6100	Consciousness lost

Effect of breathing 100% oxygen

- Delays the above effects to higher altitudes by maintaining higher arterial partial pressures at higher altitudes.
- Maximum height at which consciousness sustainable by maximal hyperventilation with 100% O_2 = 13 700 m.

Acute mountain sickness

- Occurs at levels above 3500 m.
- Due to hyperventilation causing hypocapnia and ensuing alkalosis which causes cerebral vasoconstriction.

Symptoms: headache, malaise, irritability, insomnia, poor mentation.

High-altitude pulmonary oedema

- Occurs at levels above 3500 m if ascent is rapid or heavy exercise undertaken, above 5000 m if ascent is slow.
- Due to hypoxia causing pulmonary vasoconstriction and raising pulmonary vascular resistance. This causes pulmonary oedema with increases in pulmonary blood flow such as during exercise.

Symptoms: frothy blood-stained sputum, severe dyspnoea on exercise.

3

High-altitude cerebral oedema

- Occurs at levels above 4500 m with rapid ascent, above 5000 m with gradual ascent.
- Due to progressive hypoxaemia causing progressively worsening cerebral vasodilatation, and loss of normal cerebral blood flow autoregulation. Any increase in cardiac output results in cerebral oedema. *Symptoms*: severe headache, ataxia, papilloedema, coma, fitting.

Acclimatization to altitude

1 Alteration of cerebrospinal fluid (CSF) H^+ balance. Since hyperventilation leads to hypocapnia and thus alkalosis, gradual increases in CSF carbonic anhydrase activity lead to return of the CSF H^+ concentration to normal in the presence of lower serum levels of CO_2. This reverses the inhibitory effect of hypocapnia on the respiratory centres allowing relatively more hyperventilation to occur at a given level of hypoxia.

2 Polycythaemia. Hypoxic stimulation of the juxtaglomerular apparatus of the kidney leads to increased erythropoietin levels and hence increased red cell production and polycythaemia.

3 Increased 2,3-DPG. Shifts the HbO_2 dissociation curve to the right to cause increased O_2 delivery to tissues at lower partial pressures of O_2.

4 Reduced vascular responses to hypoxia. Persistent hypoxia allows reduction in the response of vessels to local hypoxia and thus reduces the effects of hypoxia on the pulmonary and cerebral circulations. This reduces the risk of high-altitude cerebral and pulmonary oedema at given altitudes.

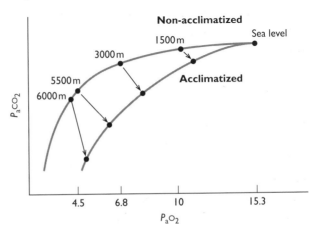

Fig. 29 Graph of P_aO_2 with altitude

Exercise

Hyperventilation ($\uparrow V_A$)

May be caused by any of the following:
- increased carotid body chemoreceptor discharge;
- increased plasma level of K^+;
- increased limb proprioceptor stimulation during exercise;
- increased cortical activity immediately prior to exercise (pre-exercise augmentation of blood O_2).

Response of carotid body may be increased not by the total blood O_2 level (which is unchanged) but by:
- increased rate of change of P_aO_2 ($\uparrow dp/dt$);
- increased rate of change of P_aCO_2 ($\uparrow dp/dt$);
- increased frequency of fluctuations in P_aO_2 (due to tachycardia causing the arterial blood to be presented to the chemoreceptors more frequently);
- similar increased frequency of fluctuations in P_aCO_2;
- increased peak and decreased trough levels of P_aO_2;
- augmentation of normal response by the fall in blood pH at very strenuous levels of exercise.

Leads to:
- increased frequency of ventilation;
- increased depth of ventilation (increased V_T).

Fall in venous oxygen concentration ($\downarrow P_vO_2$)

Caused by:
- increased exposure of capillary blood to myoglobin via muscular vasodilatation stripping out O_2 from Hb;
- increased temperature in exercising muscle;
- increased PCO_2 in muscular extracellular fluid (ECF);
- decreased muscle pH from local lactate production.

Leads to increased O_2 extraction from Hb to active muscle fibres of up to 85–90%.

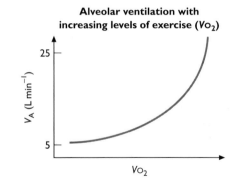

Arterial blood gases with increasing levels of exercise

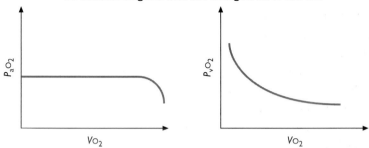

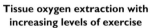

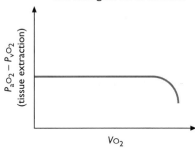

Fig. 30

Response to hypoxia and hypercapnia

Central chemoreceptors

- Lie in the ventral medullary formation.
- Primarily respond to hypercapnia (80% of hypercapnic response).
- Response:
  (a) not directly to blood gas pressures;
  (b) mainly to CSF levels of H^+;
  (c) stimulates inspiratory centre causing (i) increased ventilatory rate, (ii) increased depth of ventilation;
  (d) reduced by (i) sleep, (ii) increasing age, (iii) narcotic analgesics, (iv) alcohol, (v) athletic training, (vi) anaesthetic agents.

Peripheral chemoreceptors

- Lie:
  (a) in the carotid body at the origin of the internal carotid artery (innervation—carotid nerve branch of IX cranial nerve);
  (b) in the ascending aortic wall (innervation—aortic plexus from X cranial nerve).
- Primarily respond to hypoxia (100% of hypoxic response).
- Activity greatly augmented by hypercapnia and acidosis.
- Weakly respond directly to hypercapnia and acidosis (20% of hypercapnic response).
- Response—stimulates inspiratory centre causing:
  (a) increased ventilatory rate;
  (b) increased depth of ventilation.

Under normal circumstances (at near-normal levels of blood gas partial pressures), since the steepest response curve is that for CO_2 and not O_2, the prime effector of ventilatory control is alteration in arterial CO_2 partial pressure.

Under circumstances of hypoxia (8 kPa and less), the response curve for O_2 rises steeply, particularly when hypercapnia is also present. This makes hypoxia a progressively more potent stimulus to ventilation with worsening hypoxia.

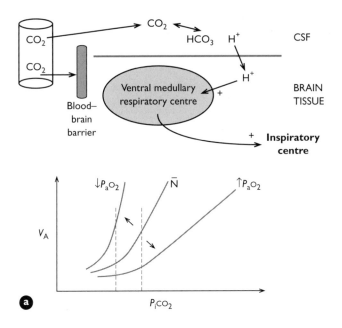

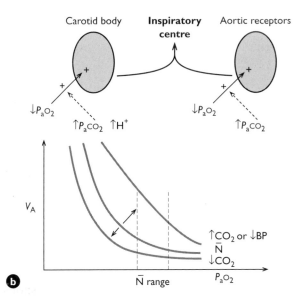

Fig. 31 (a) Central chemoreceptors; (b) peripheral chemoreceptors
BP, blood pressure; $\overline{N}$, normal.

Effects of age

Respiratory volumes

- Increased FRV.
- Decreased IRV.
- Decreased ERV.
- Increased physiological dead space.

All due to 'normal' loss of interstitial lung tissue and resultant expansion in alveolar volumes (emphysema-like change).

Gas transfer

- Reduced gas transfer area.
- Reduced diffusion capacity.

Both due to the loss in interstitial tissue causing loss of available active alveolar area. These necessitate a concomitant increase in respiratory rate and depth, particularly in exercise.

Elasticity and compliance

The loss of supportive interstitial tissue (collagen and elastin) causes an *increase* in compliance and thus a reduction in the work of breathing for each breath. This helps to offset the need for increased depth and rate of respiration occasioned by the falls in gas transfer and alterations in respiratory volumes.

Dynamic collapse

The increased compliance, however, is accompanied by increased dynamic airway collapse at progressively higher lung volumes and this leads to the approach of the collapsing volume to FRV. This eventually results in air trapping at rest. This then causes a rapid increase again in the work of breathing due to the need to open collapsed airways early in the respiratory cycle. It also reduces the alveoli available for gas transfer.

Resistance

Airflow resistance is largely unaffected until dynamic collapse begins to occur.

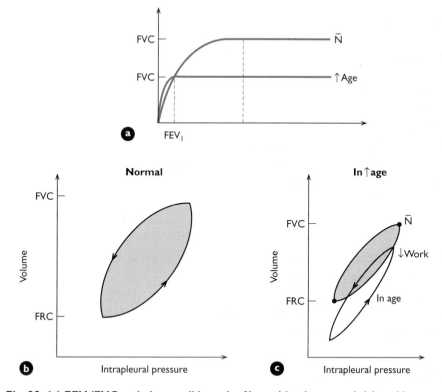

Fig. 32 (a) FEV$_1$/FVC ratio in age; (b) work of breathing in normal; (c) and in age (compared to normal)

FEV$_1$, forced expiratory volume in 1 s; FRC, functional residual capacity.

Obstructive lung disease

Asthma (obstructive disease)

Resistance

The underlying alteration to respiratory function is increased airway resistance.
Causes:
- increased bronchiolar smooth muscle tone → ↓airway radius and ∴ quadrupling of airway resistance for a halving of radius (Poiseuille–Hagen formula, see p. 19);
- increased production of mucus by bronchiolar glands → further decrease in airway diameter and increased airflow turbulence;
- chronically hypertrophied bronchial and bronchiolar epithelium → further decrease in airway radius.

Compliance

Increased bronchial epithelial thickness combined with peribronchial infiltration of inflammatory cells decreases compliance. Compliance also falls due to air trapping (see below).

Work of breathing

Increased resistance and decreased compliance increase the work of breathing both during acute episodes and at rest in chronic disease. Dynamic airway collapse significantly contributes to the work of breathing by increasing the energy required to expand collapsing airways at the start of inspiration. The air trapping resulting from failure to expand collapsed airways raises the FRV and since compliance is lower at larger volumes this further acts to increase workload.

Respiratory volumes

Increased airway resistance:
- reduces FEV_1/FVC ratio;
- increases dynamic airway collapse;
- increases FRV due to air trapping.

Gas exchange

There is no loss of gas-exchange area or functional alveoli.

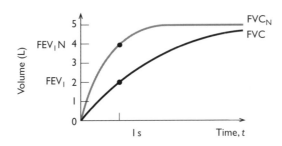

Fig. 33 FEV₁/FVC ratio

$FEV_1/FVC < 75\%$. In marked airway narrowing, FVC is also reduced but proportionally less than FEV_1. A positive bronchodilator response is a 12% increase in FEV_1 or FVC.

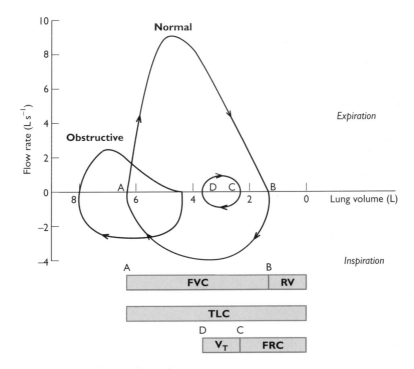

Fig. 34 Respiratory flow volume loop
FVC, forced vital capacity; RV, residual volume; V_T, tidal volume.

Restrictive lung disease

Pulmonary fibrosis (restrictive disease)

Effects

- *Compliance*. Greatly decreased due to peribronchial fibrosis and thickening of interstitial tissue by inelastic fibrous tissue.
- *Resistance*. In extensive fibrosis slight decreases in smaller airway diameter cause minor rises in resistance but it plays little role in the dynamics of respiration.
- *Work of breathing*. Greatly increased due to falls in compliance.
- *Gas transfer*. Perialveolar thickening markedly reduces the diffusion of gas across the alveolar membrane. Hypoxia in interstitial fibrosis is predominantly due to failure of gas transfer rather than disordered mechanics.
- *Airway collapse*. Normal or reduced since interstitial fibrosis splints the smaller airways preventing collapse even in forced expiration.

Respiratory volumes

- Reduced FVC.
- Reduced TLC.
- Slightly reduced FEV and ERV.
- Increased FEV_1/FVC ratio.
    At rest compensation for increased work of breathing results in:
- decreased V_T;
- increased RR.

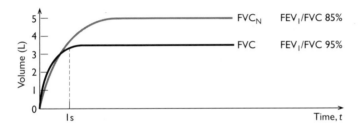

Fig. 35 FEV$_1$/FVC ratio

FEV$_1$/FVC normal or increased (FEV$_1$ and FVC are reduced in approximately equal proportions). A reduction in TLC is characteristic. Restriction cannot be diagnosed from spirometry in the presence of moderate/severe airflow obstruction.

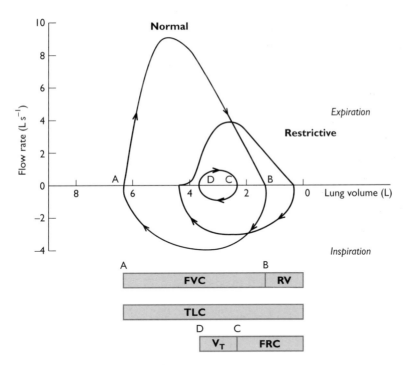

Fig. 36 Respiratory flow volume loop
RV, residual volume.

Anaesthesia and postoperative hypoxia

Significant effects on respiratory function occur after laparotomy. The causes are complex and depend not only on the direct effects of the general anaesthesia but also the effects of the co-prescribed drugs and the operation itself. These are described opposite but the most significant are:

- pain—this causes reduced respiratory excursion ($\downarrow V_T$) with a compensatory tachypnoea ($\uparrow$RR) which maintains ventilation (normal $\dot{V}$) but, since dead space remains unchanged or increases, *alveolar* ventilation is reduced ($\downarrow V_A$) to such an extent that hypoxaemia is reproducibly present in up to 95% of patients for up to 7 days post laparotomy;
- supine position (bed rest)—this causes further diaphragmatic splintage from abdominal contents (particularly if there is gastrointestinal paresis from the operative procedure causing paralytic ileus). This augments the reduction in V_T and alveolar ventilation ($\downarrow V_A$).

The role of opiate analgesics

Opiate analgesics are regarded as depressant to the central chemoreceptors, particularly in regard to their response to persistent hypercapnia. However, the role of opiates in the postoperative patient is to provide adequate analgesia, both to restore normal respiratory patterns and also to allow mobilization from the supine position and cooperation with physiotherapy to counteract the effects of secretion retention. The net effect is thus an increase in O_2 levels and reversal of hypoxaemia and hypercapnia. Opiate analgesia is thus of prime importance in counteracting the pathophysiology of the postoperative patient.

The role of inhalational general anaesthesia

Inhalational anaesthetics increase mucus secretion by a direct irritant effect on the bronchial glands and a barotrauma effect on the mucosa. These effects also reduce mucociliary escalator function. In addition, there is an uncertain effect on smaller airways promoting dynamic collapse by causing the dynamic collapsing volume to approach FRV.

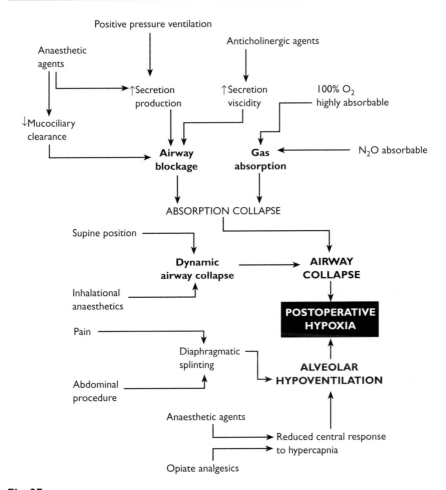

Fig. 37

3

Pulmonary embolism

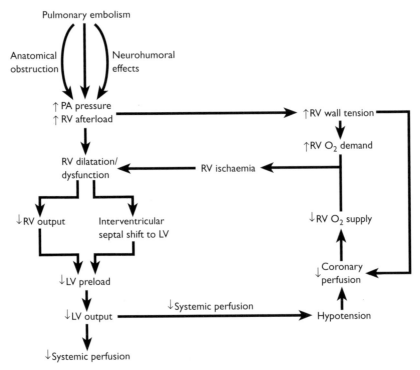

Fig. 38

PA, pulmonary artery. (From Fauci, A., Braunwald, E., Isselbacher, K.J. *et al.* (eds) (1998) *Harrison's Principles of Internal Medicine.* McGraw-Hill, New York.)

Pulmonary embolism results in:
- increased pulmonary vascular resistance due to vascular obstruction and neurohumoral factors;
- pulmonary hypertension increasing RV afterload, increasing RV wall tension, reducing right coronary artery perfusion and resulting in RV dilatation and dysfunction;
- reduced LV output due to ventricular interdependency, reduced coronary perfusion and a fall in LV preload;
- impaired gas exchange due to increased alveolar dead space, low *V/Q* units and right-to-left shunting;
- decreased lung compliance due to pulmonary oedema, pulmonary haemorrhage and loss of surfactant.

Pneumonectomy

Effects at rest

- Decreased FRV.
- Decreased FVC.
- Decreased IRV.
- Decreased ERV.
- Decreased anatomical and physiological dead space.

All due to the decrease in overall number of alveoli available and reduced volume of conducting airways.

- Increased resting RR.
- Increased resting V_T.

These compensate for the reduction in the resting total available gas-exchange area and lead to:

- no change in pulmonary artery pressures;
- no change in arterial Po_2;
- no change in red cell count or Hb concentration.

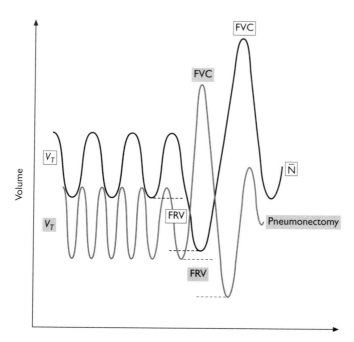

Fig. 39 Respiratory volumes after pneumonectomy

4: RENAL PHYSIOLOGY

4

Basic principles

Clearance is the volume of plasma completely cleared of solute by the kidney in 1 min.

Clearance ratio for a substance indicates degree of active excretion or reabsorption (ratio > 1 = actively excreted; < 1 = actively reabsorbed).

$$\text{Clearance} (Cl_s) = \frac{[U]_s}{[P]_s} \times V$$

Glomerular filtration rate (GFR) = Cl_s for any substance not metabolized by the kidney, actively excreted by the tubules or actively reabsorbed by the tubules.

$$\text{GFR} = \frac{[U]_s \times V}{[P]_s} = 125 \, \text{mL min}^{-1}$$

$$\text{Renal plasma flow (RPF)} = \frac{[U]_s \times V}{[P_a]_s - [P_v]_s} = 600\text{--}700 \, \text{mL min}^{-1}$$

$$\text{Renal blood flow (RBF)} = \frac{\text{RPF}}{(1 - \text{haematocrit})} = 1200 \, \text{mL min}^{-1}$$

$$\text{Filtration fraction} = \frac{\text{GFR}}{\text{RPF}} = 0.25$$

$$\text{Clearance ratio} = \frac{Cl_s}{\text{GFR}}$$

$$\text{Creatinine clearance} = \frac{[U]_{Cr} \times V}{[P]_{Cr}} \approx \frac{88\,(145 - \text{age})}{[S]_{Cr}} - 3$$

where $[U]_s$ is urine concentration of substance s; $[P]_s$ is plasma concentration of substance s (a is arterial, v is venous); and V is urine flow rate.

Factors determining glomerular filtration rate

- Regulation of RBF:
  (a) sympathetic activity (decreases RBF and therefore GFR);
  (b) endocrine and paracrine mechanisms (i) endothelins (decrease RBF and GFR), (ii) angiotensin II (constricts efferent arterioles more than

afferent arterioles, thereby increasing glomerular hydrostatic pressure and GFR), (iii) nitric oxide (NO) (increases RBF and GFR);

(c) autoregulation (i) tubuloglomerular feedback, (ii) intrinsic myogenic mechanism (constriction when blood pressure rises).

• Hydrostatic pressure gradient across capillary wall (determined by arterial pressure, afferent and efferent arteriolar resistance).

• Osmotic pressure gradient across capillary wall.

• Capillary permeability (50 times greater than that of skeletal muscle).

• Size of the capillary bed (determined by mesangial cell contraction and relaxation).

• Electrical charge (sialoproteins in the glomerular capillary wall are negatively charged; circulating albumin is negatively charged and therefore not filtered easily).

Measurement of GFR. Any substance that is filtered but not reabsorbed, actively excreted by the tubules or metabolized by the tubular epithelium will have a clearance which represents only the GFR.

• Creatinine:
 (a) endogenous substance;
 (b) easily measured serum level without intervention;
 (c) slightly actively excreted;
 (d) test slightly over-measures urine creatinine (approximately balances out effect of (c));
 (e) affected by starvation and exercise.

• Inulin:
 (a) exogenous substance;
 (b) needs infusion to achieve serum level;
 (c) highly accurate reflection of GFR.

Measurement of RPF. Any substance completely removed from the renal arterial blood by the kidney has $[P_v] = 0$. Therefore $Cl_s = \text{RPF}$.

• *Para*-aminohippuric acid (PAH):
 (a) highly removed by kidney but not 100%;
 (b) blood flow to non-cortical tissue not accounted for;
 (c) measures *effective* RPF.

Body fluid compositions

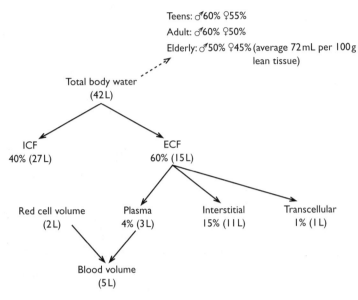

Fig. 40 Typical values
Figures in parentheses are typical values for a 70 kg, 1.82 m male. ECF, extracellular fluid; ICF, intracellular fluid.

Measurement of compartments

- Total body water (TBW): deuterium dilution.
- ECF: inulin or thiosulphate dilution.
- Plasma: protein-bound dye dilution (e.g. Evans blue/iodinated albumin).
- Interstitial: ECF – plasma.
- Intercellular ICF: TBW – ECF.

Typical body fluid values (means/mmol L^{-1})

	Plasma	ECF	ICF	Lymph
Na$^+$	142	145	12	140
K$^+$	4.3	4.4	150	
Ca^{2+}	2.5	2.4	4.0	
Mg^{2+}	1.1	1.1	34	
Cl$^-$	104	117	4.0	
PO$_4^{2-}$	2.0	2.0	40	
Prot$^-$	14	0	54	
HCO$_3^-$	24	27	12	

Typical daily fluid and electrolyte balance

		H_2O		Na$^+$ (mmol)	K$^+$ (mmol)	Ca^{2+} (mmol)	Mg^{2+} (mmol)
In	1200 drink	800 food	400 metabolism	160	100	950	300
	↓	↓	↓	↓	↓	↓	↓
Out	1400 urine	800 respiration and sweat	200 faeces	155	95	945	290

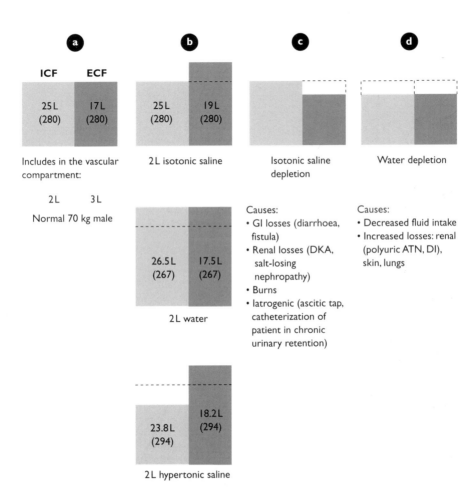

Fig. 41 (a) Fluid distribution in normal adult. (b) Effects of different fluid loads on fluid distribution and osmolality. (c) Saline depletion. (d) Water depletion
Numbers in parentheses represent osmolality (mOsm kg^{-1}). The dashed lines represent normal limits. ATN, acute tubular necrosis; DI, diabetes insipidus; DKA, diabetic ketoacidosis; GI, gastrointestinal.

Water balance

Antidiuretic hormone (ADH) (also known as arginine vasopressin)

- Released from supraoptic hypothalamic nuclei.
- Stimuli to release:
  (a) plasma hyperosmolality;
  (b) nicotinic drugs;
  (c) opioids.
- Actions on the kidney:
  (a) increased collecting duct epithelial permeability to water;
  (b) increased collecting duct permeability to urea (allows loss to lumen);
  (c) decreased vasa recta blood flow (decreases medullary solute gradient generation);
  (d) possible action to decrease distal convoluted tubule Na^+ reuptake.
- Actions on other tissues: vasoconstrictor—primarily arteriolar (not significant in physiological output range but output is greatly enhanced by hypotension and greater levels play a significant role in the response to haemorrhage).
- Degradation:
  (a) $t_{1/2} = 10$ min;
  (b) 80% metabolized in liver;
  (c) 20% metabolized in kidney or excreted.

Syndrome of inappropriate antidiuretic hormone secretion (SIADH)

Diagnostic criteria

- Plasma Na < 130 mmol L^{-1}, urinary Na > 20 mmol L^{-1}.
- Urine osmolality > 500, urine osmolality > plasma osmolality (usually < 275 in SIADH).
- Normovolaemia, normal renal and endocrine function.
- Off diuretics.

Causes

- Malignancy (lung, pancreas, prostate).
- Central nervous system (meningoencephalitis, cerebrovascular accident, subarachnoid haemorrhage).
- Chest disease (tuberculosis, pneumonia, abscess).
- Drugs (opiates, carbamazepine, thiazides).

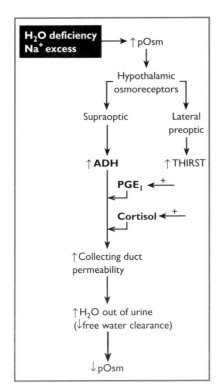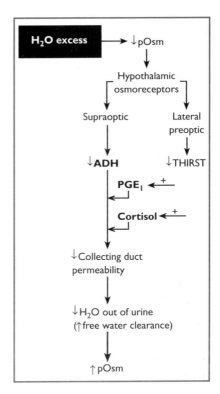

Fig. 42
PGE$_1$, prostaglandin E$_1$; pOsm, plasma osmolality.

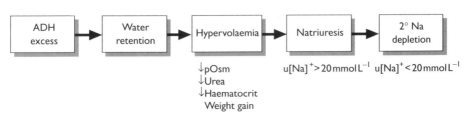

Fig. 43
u[Na]$^+$, urinary sodium concentration.

Sodium balance

Sodium and the relation to volume

Since sodium is principally an extracellular ion, total body level of sodium (reflected in ECF concentration) helps determine body water distribution. Sodium retention is accompanied by increase in osmolality with resultant increased free water retention. Thus, sodium is a principal method of determining total body water. Equally, the response to changes in volume (due to haemorrhage, etc.) closely parallels the response to hypernatraemia.

Osmolality = moles per kg of water = 2Na + urea + glucose

(Note: osmolarity = moles per litre of solution.)
A discrepancy between measured and calculated plasma osmolality occurs in the presence of other osmotically active substances:
- hyperlipidaemia;
- hyperproteinaemia;
- hyperglycaemia;
- other: ethanol, methanol, ethylene glycol, mannitol, salicylate, glycine (used in irrigation fluids during transurethral resection of prostate).

Renin–angiotensin system

Angiotensinogen

- Liver-produced α_2 globulin.
- Release also stimulated by sympathetic response.

Angiotensins

- Rapidly degraded by liver enzymes.
- Act primarily via effect on aldosterone release.
- May also have direct antinatriuretic effect on DCT.

Aldosterone

- Effect via nuclear receptor binding and alteration of DNA production to effect increased production of cellular membrane pump proteins.
- Effect is primarily permissive and augmentative in pathological states.

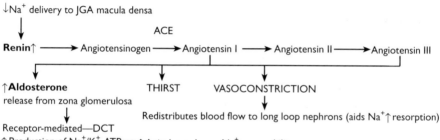

Fig. 44
ACE, angiotensin-converting enzyme; JGA, juxtaglomerular apparatus.

4

Renal sympathetic nerves

- Stimulated primarily by hypotensive response.
- Augment aldosterone activity on DCT epithelium.
- Augment hyponatraemic renin release.

Atrial natriuretic peptide

Atrial natriuretic peptide (ANP) occurs predominantly in myocardium (atrium) and acts principally on the natriuretic peptide receptor A (NPR-A).

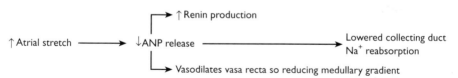

Fig. 45

Other natriuretic peptides

- Brain natriuretic peptide (BNP) also occurs predominantly in the myocardium (ventricle > atrium).
- C-type natriuretic peptide (CNP) occurs principally in the brain and acts on natriuretic peptide receptor B (NPR-B). CNP is not natriuretic but possesses vasorelaxant properties.

Disorders of sodium balance

Causes of hyponatraemia	Causes of hypernatraemia
Sodium loss	*Sodium retention*
Renal:	Primary
Diuretic excess	Hyperaldosteronism
Mineralocorticoid deficiency	Cushing's syndrome
Salt-losing nephropathy	Hypertonic dialysis
Bicarbonaturia with RTA and metabolic alkalosis	Hypertonic Na HCO_3
Ketonuria	NaCl tablets
Osmotic diuresis	
	Water loss
Extra-renal:	Renal:
Vomiting	Osmotic or loop diuretic
Diarrhoea	Postobstruction
Third spacing of fluids: burns, pancreatitis, trauma	Intrinsic renal disease
	Diabetes insipidus
Excess body water	Hypodipsia
Glucocorticoid deficiency	
Hypothyroidism	Extra-renal:
Stress	Excess sweating: burns, diarrhoea, fistulas
Drugs	Insensible losses
SIADH	Respiratory
Acute or chronic renal failure	Dermal
Nephrotic syndrome	
Cirrhosis	
Cardiac failure	

RTA, renal tubular acidosis.

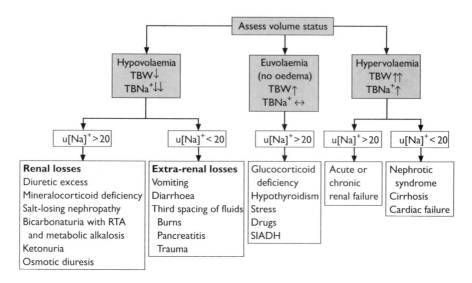

Fig. 46 Diagnostic algorithm for hypo-osmolar hyponatraemia
TBNa+, total body Na+; TBW, total body water.

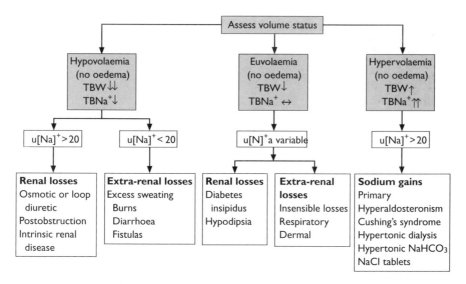

Fig. 47 Diagnostic algorithm for hypernatraemia
(Figures 46 & 47 adapted from Halterman, R. & Berl, T. (1998) Therapy of dysnatremic disorders. In: *Therapy in Nephrology and Hypertension* (eds Brady, H. & Wilcox, C.). W.B. Saunders, Philadelphia.)

Potassium balance

Causes of hypokalaemia	Causes of hyperkalaemia
Shifts into cells:	Shifts out of cells:
Systemic alkalosis	Burns
Insulin infusion	Trauma/rhabdomyolysis
Renal loss:	Acidosis
Diuretics	Renal retention:
Hyperaldosteronism	ARF
Cushing's syndrome (including	CRF
glucocorticoid therapy)	Addison's disease
RTA	K^+-sparing diuretics
Renal drug toxicity	
Gastrointestinal loss:	
Diarrhoea	
Vomiting	
Stomas (especially ileostomy)	
Secretory diarrhoea (e.g. rectal adenomas, ulcerative colitis)	

ARF, acute renal failure; CRF, chronic renal failure.

Mechanisms of correction

Distal convoluted tubular handling

- Normally handles 15–20% of filtered K^+ load.
- Driven by simple electrochemical gradient.
- Higher serum level increases gradient to filter more K^+.

Aldosterone

- From zona glomerulosa cells of adrenal cortex.
- Secretion fairly rapid in response to raised $[K^+]$ in serum flowing in adrenal cortical blood.
- Action to increase both number and activity of two K^+ handling pumps on the apical membrane of the DCT epithelium.
- Increases Na^+ reabsorption by up to 2% (in association with K^+ and H^+ secretion and Cl^- reabsorption) by:
    (a) increasing absolute numbers of ENaC (specific epithelial amiloride-sensitive Na^+ channels, also blocked by trimethoprim) on the luminal surface of cortical collecting duct (CCD);
    (b) encouraging ENaC to remain in an open configuration;
    (c) increasing the number of Na^+ pumps.

Primary hyperaldosteronism

• Aldosterone overproduction independent of regulation by AII results in salt and water retention. This results in suppression of renin and angiotensin secretion. Other features include increased urinary K^+ loss (> 20 mmol day^{-1}), hypokalaemia, alkalosis and hypertension.
• Most commonly due to aldosterone-producing adrenocortical adenomas (sporadic or familial) or bilateral adrenal cortical hyperplasia.
• Plasma aldosterone–renin activity in a random sample taken from a patient seated for 5 min is a sensitive screening test.
• Suppression of aldosterone by saline infusion (2 L over 4 h) or following fludrocortisone pretreatment excludes the diagnosis.

Cellular ion shifts

• Occur almost immediately in response to raised ECF $[K^+]$.
• Occur due to passive diffusion gradient.
• Electrochemical gradient causes H^+ (available intracellular cation) to flow out of the cells.
• Do not correct underlying total body 'hyperkalosis' but correct immediate hyperkalaemia.
• Once hyperkalaemia corrected subsequent reverse shift occurs (leading to potential of so-called 'afterdrop' of H^+).

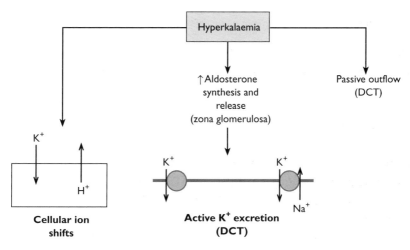

Fig. 48

Acid–base balance I

Principles

- pH is the negative log of H^+ concentration.
- Buffers are most effective at influencing pH over ± 1 pH point of their pK value.
- The main affector ECF buffer is $H_2CO_3/HCO_3^-/CO_2$.
- The HCO_3^-/CO_2 buffer is unique in that CO_2 is controlled by ventilation giving a variable effective pK.
- pH is thus influenced by HCO_3^- concentration (determined by renal function) and P_aCO_2 (determined by respiratory function).

Key terms

- *Base excess*: the amount of base required to restore 1 L of the blood sampled to a pH of 7.0 at $P_{CO_2} = 5.3$ kPa.
- *Anion gap*: the amount of anions other than Cl^- and HCO_3^- which electroneutralize Na^+ ions. It is essentially made up of $H_2PO_4^-$ and SO_4^{2-} (less Ca^{2+} and Mg^{2+}). It is increased in those forms of metabolic acidosis that produce more anions (e.g. lactate and ketoacids). Anion gap $= Na - (Cl + HCO_3) = 15–20$ mmol L^{-1}. Albumin accounts for 10 mmol L^{-1} of anion gap.

Temporary serum buffers

Note. The bulk of acid load is buffered by intracellular $H_2PO_4^-$ and HCO_3^- and haemoglobin (Hb). Additional extracellular buffers are albumin and $H_2PO_4^-$.

Features: immediate, moderate capacity, temporary.

Buffer	pK value	Location	Effective buffer capacity (relates to concentration pK_{eff})
HCO_3^-	6.1	Plasma ECF	18
HPO_4^{2-}	6.8	ICF	0.3
Hb	Varies	Plasma	8
Protein (NH groups)	Varies	Plasma	Varies

Respiratory buffer

Features: very rapid, large capacity, temporary, only available if no lung disease.

$$CO_2 + H_2O \Leftrightarrow H_2CO_3 \Leftrightarrow H^+ + HCO_3^-$$
$$\downarrow$$

removed by hyperventilation

CO_2/HCO_3^- normally 1/20

4

Cellular shifts

Features: hours onset, moderate capacity, temporary.

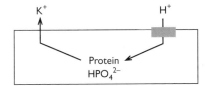

Fig. 49

Renal correction

Features: hours–days onset, infinite capacity.

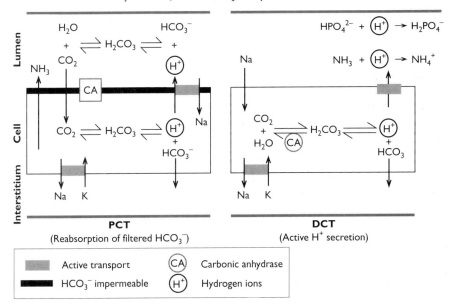

Fig. 50 Role of kidney in acid–base balance
PCT, proximal convoluted tubule.

Acid–base balance 2

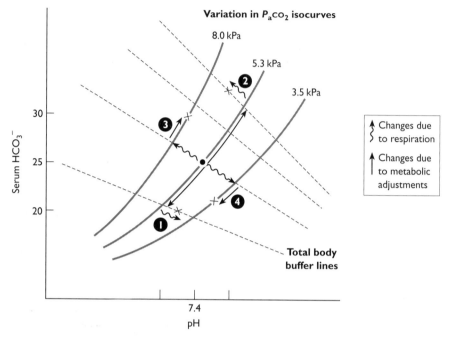

Fig. 51

Notes

- Metabolic shifts/renal compensation shifts can be thought of as occurring along CO_2 isobars initially with a subsequent respiratory compensation.
- Metabolic/renal changes change the standard bicarbonate and base excess due to actual changes in body bicarbonate or acid levels.
- Respiratory changes occur along total body buffer lines with subsequent (slower) metabolic compensatory changes.
- Respiratory changes do not alter standard bicarbonate or base excess since changes to bicarbonate concentration in the plasma are only due to the effects of the $CO_2/H_2CO_3/HCO_3^-$ buffer system and vanish when the bicarbonate is measured in the *normal or standard* situation in the laboratory.

Metabolic acidosis effects ❶

	Initial insult	After acute respiratory compensation	After renal compensation
pH	$\Downarrow$	$\downarrow$	–
CO_2	$\leftrightarrow$	$\Downarrow$	–
HCO_3^-	$\downarrow$	$\Downarrow$	Secrete H^+/retain HCO_3^-
St Bicarb	Low	Low	Normal
Base Exs	−ve	−ve	0

Metabolic alkalosis effects ❷

	Initial insult	After acute respiratory compensation	After renal compensation
pH	$\Uparrow$	$\uparrow$	–
CO_2	$\leftrightarrow$	$\Uparrow$	–
HCO_3^-	$\uparrow$	$\Uparrow$	Retain H^+/secrete HCO_3^-
St Bicarb	High	High	Normal
Base Exs	+ve	+ve	0

Respiratory acidosis effects ❸

	Initial insult	After chronic metabolic compensation
pH	$\Downarrow$	$\downarrow$
P_aCO_2	$\Uparrow$	$\Uparrow$
HCO_3^-	$\uparrow$	$\Uparrow$
St Bicarb	Normal	High
Base Exs	0	+ve

Respiratory alkalosis effects ❹

	Initial insult	After chronic metabolic compensation
pH	$\Uparrow$	$\uparrow$
P_aCO_2	$\Downarrow$	$\Downarrow$
HCO_3^-	$\downarrow$	$\Downarrow$
St Bicarb	Normal	Low
Base Exs	0	−ve

$\Uparrow/\Downarrow$, large increase/decrease.
$\uparrow/\downarrow$, small increase/decrease.
$\leftrightarrow$, no change.

Acid–base balance 3

Causes

Metabolic acidosis

High anion gap	Normal anion gap
• Uraemic acidosis	• GIT HCO_3 loss—diarrhoea, pancreatic fistula
• Diabetic ketoacidosis	• RTA (see opposite)
• Lactic acidosis	• Dilutional
• Starvation ketoacidosis	• Infusion of TPN (amino acids)
• Drugs (salicylates)	

GIT, gastrointestinal tract; TPN, total parenteral nutrition.

Metabolic alkalosis

- Vomiting.
- Hyperaldosteronism (Conn's syndrome, reninoma, carbenoxolone).
- Alkali intoxication (ingestion, infusion).
- Diuretic therapy (K^+, Na^+ loss).

Respiratory acidosis

- Lung parenchymal disease.
- Mechanical ventilatory failure.
- Neuromuscular disorders.
- Central nervous system depression (drugs, head injury).
- Abdominal pain/surgery.

Respiratory alkalosis

- Hysterical hyperventilation.
- Anaesthetic hyperventilation (e.g. intensive therapy unit).
- Pain.

Renal tubular acidosis

In RTA, there is a failure to acidify urine appropriately. Systemic acidosis is not corrected. SO_4^{2-} and PO_4^{2-} ions are filtered normally and electrical neutrality must be maintained by renal Cl^- reabsorption. This results in a hyperchloraemic metabolic acidosis with a normal anion gap.

Proximal and distal RTAs are associated with K^+ loss.

4

Proximal (type 2) RTA	Distal (type 1) RTA	Type 4 RTA
• Proximal HCO_3 reabsorption is impaired resulting in HCO_3 wasting • Urinary pH is 4.5	• This results from a failure to secrete H^+ ions in the distal tubule • Patients cannot reduce urinary pH below 5.5	• This results from defective NH_3 production • Quantity of acid excreted is reduced because of insufficient urinary buffer
Causes • Wilson's disease • Cystinosis • Galactosaemia • Autoimmune (Sjögren's) • Renal (myeloma, nephrotic syndrome, transplant) • Drugs (acetazolamide, lead, tetracyclines)	Causes • Primary • Secondary (a) autoimmune (Sjögren's syndrome, PBC) (b) renal (nephrocalcinosis, transplant) (c) drugs (amphotericin, lithium, non-steroidal anti-inflammatory drugs)	Causes • Hyperkalaemia • Glucocorticoid deficiency • Mineralocorticoid deficiency (hyperkalaemia suppresses NH_3 production)

PBC, primary biliary cirrhosis.

Alkali therapy in acid–base disturbances

Alkali therapy (HCO_3 in particular) carries a risk of pH-mediated depression of ventilation, enhanced CO_2 production from HCO_3 decomposition (and thus paradoxical worsening of intracellular acidosis) and volume expansion.

Renal tubular function: summary

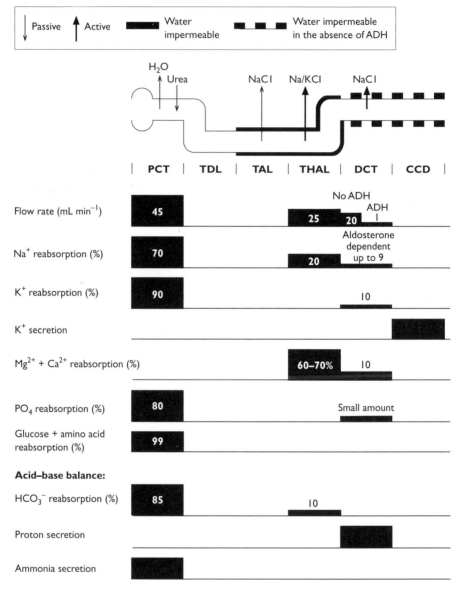

Fig. 52 Summary of renal tubular function: regional differences
TAL, thin ascending limb; TDL, thin descending limb; THAL, thick ascending limb.

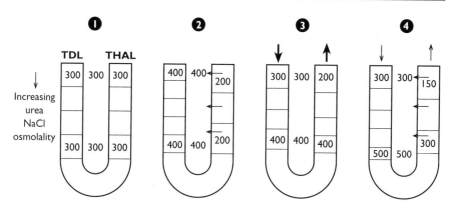

Fig. 53 Mechanism of countercurrent multiplication

Numbers represent urine osmolality (mOsm kg^{-1}). Spacing of lines within tubule is equivalent to osmolality. Thin arrows indicate direction of active NaCl transport. Bold arrows indicate direction of urine flow.

4

5: ALIMENTARY PHYSIOLOGY

5

Gut hormones

Hormone	Actions
Somatostatin (Growth hormone-inhibiting hormone) Produced by D cells of antrum to colon Acts via somatostatin receptors (five subtypes) with seven membrane-spanning domains present in the brain, anterior pituitary, endocrine and exocrine pancreas, GI mucosa and immune cells	Reduces liver and splanchnic blood flow Inhibits gallbladder (GB) contractility and bile flow Inhibits gastric acid and pepsin secretion Inhibits gastrointestinal (GI) and pancreatic (endocrine and exocrine) hormone secretion Slows GI transit time and inhibits glucose and amino acid absorption Stimulates water and electrolyte absorption *Non-gastrointestinal effects* Inhibits tissue growth and proliferation Inhibits immune cell activity Inhibits growth hormone and thyrotrophin secretion
Gastrin Secreted by G17 (G cells in antrum) and G34 (duodenum) in response to: • vagal stimulation • food entry Acts via gastrin and cholecystokinin receptors	Stimulates gastric acid secretion by parietal cells Stimulates pepsinogen release by chief cells Stimulates histamine release by enterochromaffin-like cells (ECL) Stimulates somatostatin release (negative feedback) Relaxes pyloric sphincter Increases lower oesophageal sphincter pressure Stimulates pancreatic enzyme secretion Promotes gastric and intestinal motility and secretion
Cholecystokinin (CCK) Produced by I cells in duodenum and proximal jejunum in response to intraluminal fat, amino acids, peptides and cations (Ca^{2+} and Mg^{2+})	Stimulates GB contraction Stimulates pancreatic enzyme and HCO_3 release (potentiates secretin action) Delays gastric emptying Inhibits small intestine motility Modifies eating behaviour
Secretin Produced by S cells in villi and crypts of small intestine in response to acidification of duodenal contents (pH < 4.5)	Stimulates pancreatic enzyme and HCO_3 release (potentiates CCK action) Inhibits gastric acid and pepsin secretion
Vasoactive intestinal peptide (VIP) Secreted by small intestine in response to vagal stimulation	Evokes gastric and GB relaxation (as does substance P) Increases pancreatic and intestinal secretion Inhibits gastric acid and pepsin secretion Neurotransmitter Vasodilatation
Substance P	Evokes gastric and GB relaxation Neurotransmitter Vasodilatation Diuresis and natriuresis

Continued

Hormone	Actions
Gastric inhibitory peptide (GIP) Secreted by duodenum and jejunum	Inhibits gastric acid secretion and motility Stimulates postprandial insulin secretion
Gastrin-releasing peptide (bombesin) Widespread distribution	Stimulates gastrin and gastric acid secretion Neurotransmitter Increases pancreatic secretion
Enteroglucagon (Proglucagon fragments) Ileum and colon	Inhibits gastric acid secretion Oxyntomodulin stimulates liver glycogenolysis and insulin release

5

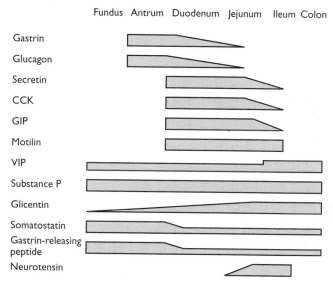

Fig. 54 Distribution of gastrointestinal peptides along the gastrointestinal tract
The thickness of each bar is proportionate to the concentration of the peptide in the mucosa.
(From Ganong, W.F. (1999) *Review of Medical Physiology*, 19th edn. Appleton & Lange,
Norwalk, CT; reproduced with permission of The McGraw-Hill Companies.)

Enteric nervous system

Structure

Nerve cells are grouped into small ganglia connected by nerve bundles forming two plexuses.

Myenteric (Auerbach's) plexus	Submucous (Meissner's) plexus
• Along entire length of gut • Lies between longitudinal and circular layers of muscle • Provides motor innervation to two muscle layers • Secretomotor innervation to mucosa • Also present in striated muscle portion of oesophagus • Innervates motor end plates with inhibitory nitric oxide (NO)	• Located in submucosa between circular muscle layer and muscularis mucosa • Best developed in the submucosa • Innervates glandular epithelium, muscularis mucosa, submucosal blood vessels, intestinal endocrine cells

Peristalsis

Peristalsis is the combination of a number of local reflexes.

• Mucosal stimulation and/or distension of the gut lumen releases 5-hydroxytryptamine (5HT).

• This produces circular muscle contraction above the stimulus via ascending cholinergic interneurons to excitatory motor neurons (containing acetylcholine (ACh) and substance P).

• Below the stimulus, descending cholinergic interneurons activate inhibitory motor neurons (containing NO, VIP and ATP) and cause muscle relaxation.

• This propels the bolus of food forward.

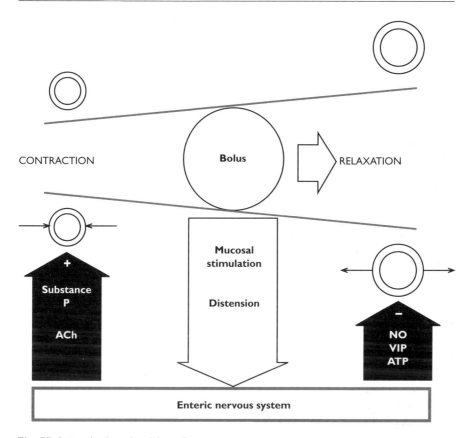

CONTRACTION

Bolus

RELAXATION

+

Substance
P

ACh

Mucosal
stimulation

Distension

–

NO
VIP
ATP

Enteric nervous system

5

Fig. 55 Intestinal peristaltic reflex

Stomach I: motility, the vagus and vagotomy

Actions of the vagus

Motor

- Direct ACh release causes stimulation of gastric oxyntic cells.
- ACh release causes stimulation of histaminocytes in gastric mucosa.
- Augmentation of gastrin release from G cells of antrum.
- Modulation of myenteric activity and hence enhanced gastric motility.
- Modulation of pyloric tone (permits episodic relaxation once stomach distended).
- Promotion of pancreatic enzyme release stimulated by CCK.

Sensory

- Sensation of gastric distension ('fullness').
- Sensation of gastric overdistension (nausea and discomfort).

Other effectors of gastric motility

Promoters of gastric peristalsis	Inhibitors of gastric peristalsis
• GI hormones: gastrin, motilin	• GI hormones: GIP, VIP, CCK, secretin
• Distension	• Overdistension and duodenal distension
• Simple carbohydrates	• Fats
• ACh	• Sympathetic α agonists (fear, etc.)
Promoters of pyloric relaxation	**Inhibitors of pyloric relaxation**
• Vagal tone (ACh)	• Sympathetic tone (α agonists)
• Gastrin	• Secretin, CCK

Effects of vagotomy

- Reduced gastric acid secretion through both reduced direct oxyntic cell stimulation and reduced mucosal histamine and gastrin release (p. 104).
- Reduced coordination of myenteric activity may lead either to delayed emptying and stasis or to early hyperactive emptying particularly if a gastric drainage procedure is present.
- Failure of pylorus to relax prior to peristaltic emptying wave.
- Reduced pancreatic exocrine (enzymatic) function.

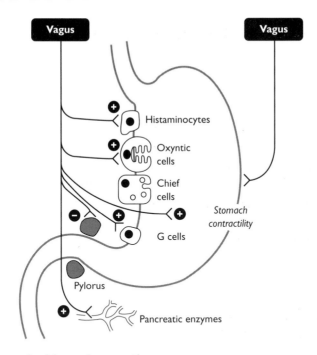

Fig. 56 Stomach with vagal connections

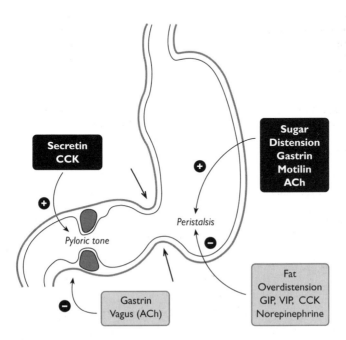

Fig. 57 Stomach motility effectors

Stomach 2: acid secretion and hypergastrinaemia

Oxyntic (parietal) cell function

- Located in the body of the stomach (the area below the lowest point of the gastro-oesophageal junction to the incisura angularis).
- Secrete H^+ and intrinsic factor (p. 110).
- H^+ secretion is by active pump-mediated membrane transport.
- H^+ output depends on overall flow rate.
- Stimulated by ACh, gastrin, histamine.
- Suppressed by $\uparrow H^+$ (luminal), secretin, GIP.

G (antral) cell function

- Located in the antrum and prepyloric area.
- Secrete gastrin which promotes oxyntic cell function, pepsinogen release (chief cells), gastric motility and pyloric opening. Gastrin also promotes intestinal motility and secretion.
- Stimulated by ACh (vagus), local stretch reflex, Ca^{2+}.
- Suppressed by $\uparrow H^+$, secretin, GIP, VIP.

Hypergastrinaemia

Causes

- G-cell tumour: type I Zollinger–Ellison syndrome.
- G-cell hyperplasia: type II Zollinger–Ellison syndrome.
- Excluded antrum after surgery (G cells uninhibited by H^+).
- Atrophic gastritis (no H^+ to suppress G cells).
- Hyperparathyroidism (excessive Ca^{2+} stimulation).
- Chronic liver disease (decreased metabolism of gastrin).

Clinical features of Zollinger–Ellison syndrome

- Fulminant, atypical, recurrent or multiple peptic ulcers.
- Watery, intermittent ('secretory') diarrhoea. Caused by episodic intestinal motility and secretion overstimulation. (Associated with proven serum hypergastrinaemia.)

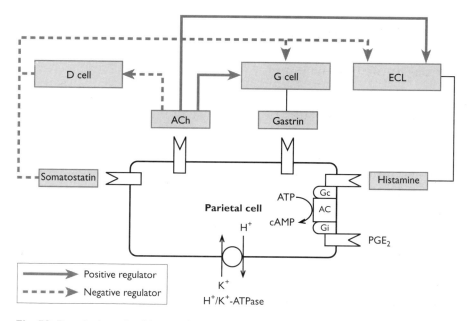

Fig. 58 Regulation of acid secretion by parietal cell
AC, adenylyl cyclase; Gc, Gi, G proteins; PGE₂, prostaglandin E₂.

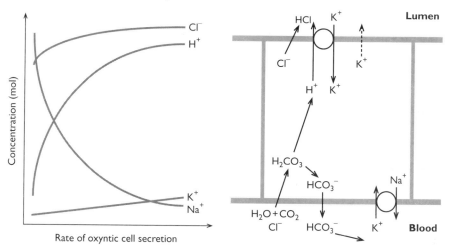

Fig. 59 Relationship of acid output to secretion rate

Fig. 60 Apical membrane pump activated during high secretion rates

Stomach 3: functions and postgastrectomy syndromes

Functions

1 Storage of food: occurs in the body and antrum. Acts to convert large-volume intermittent 'doses' of food to semiconstant moderate-volume 'doses' of small-bowel chyme.

2 Digestion of protein. Brought about by the action of pepsinogen (released from chief cells in the body) which is cleaved in the presence of H^+ to pepsin I, II or III. Pepsin cleaves phenylalanine- and tyrosine-related bonds in proteins.

3 Gastrin production: occurs from the G cells of the antrum (p. 108).

4 Intrinsic factor production: occurs from the oxyntic (parietal) cells as a result of gastrin stimulation. Output exceeds daily requirement for vitamin B_{12} by 30 times. Binds B_{12} in the gastric lumen for receptor-mediated uptake in the terminal ileum.

5 Conversion of Fe^{3+} (ferric) to Fe^{2+} (ferrous) to enable ileal receptor-mediated uptake.

6 Sterilization of gastric contents. Provides some protection against non-coated bacterial organisms ingested.

7 Absorption of food. Limited to small amounts of lipid-soluble simple compounds (e.g. alcohol).

Postgastrectomy syndromes

Can occur after any gastrectomy, partial or total.

Bilious vomiting

Due to loss of pylorus and body of stomach which act to coordinate forward propulsion of food and prevent reflux of duodenal contents into the stomach (see **1** above).

Diarrhoea

Due either to rapid transit of large volumes of osmotically active foodstuffs into the small bowel resulting in the creation of a large volume of liquid chyme, or to disordered small-intestinal motility consequent upon disruption of the vagal innervation, or incomplete commencement of digestion (see **1, 2** above).

Dumping syndrome

Characterized by faintness, tachycardia, sweating and abdominal discomfort (sympathetic nervous system activation).
• Due either to excessive volumes of osmotically active foodstuffs being liberated into the small bowel too quickly causing a transient hypovolaemia because of the passage of large volumes of extracellular fluid into the lumen of the intestine;
• or to rapid swings in serum glucose because of uncontrolled carbohydrate absorption caused by uncontrolled passage out of the stomach (see **1** above).

Small stomach syndrome

Due to loss of capacity causing premature sensations of 'fullness' leading to a risk of inadequate calorific intake.

Anaemia

• Due either to iron deficiency resulting from available iron being in the wrong ionic state for absorption (see **5** above);
• or to vitamin B_{12} deficiency because of inadequate absorption (see **4** above).

Bile secretion and the effects of cholecystectomy

Bile

- Approximately 5000 mL produced by the liver daily.
- Contains 97% water, 1.8% bicarbonate, 0.7% bile salts, 0.2% glucuronidated bilirubin/biliverdin, 0.3% cholesterol, fatty acids and lecithin.
- Bile salts:
    (a) 500 mg produced per day;
    (b) 90% recycled by terminal ileum.
- Stimulated by bile salts, secretin, CCK, vagus (ACh).
- Suppressed by fasting, sympathetic α agonists.

Gallbladder functions and the effects of cholecystectomy

Function	Postcholecystectomy
Normal concentration of bile from 5 L to 500 mL per day due to gradual water extraction by the mucosa lining the GB	Increased volume flow leading to chronic mildly raised common bile duct pressure and increased cholesterol transmetabolism to replace losses. Increased volume of bile resulting may lead to biliary reflux into the stomach with associated biliary gastritis
Production of micelles (conglomerations of bile salts, cholesterol, lecithin and salts) to aid fat absorption. Occurs due to the mixing allowed in the GB during collection of bile between meals and helps to retain the iso-osmotic nature of the bile despite overall concentration	Reduced micelle formation due to lack of time to form and may lead to fat intolerance and malabsorption with resultant abdominal cramps and diarrhoea after fatty meals. Tends to reduce with time secondary to common bile duct dilatation and stasis which permit micelle formation
Conversion of a continuous bile flow into an intermittent ejection of bile in response to CCK stimulation	Reduced maximal bile flow after large meals, particularly initially, with resultant exacerbation of fat intolerance. Tends to reduce with time due to increased common bile duct capacity and contractility in response to CCK

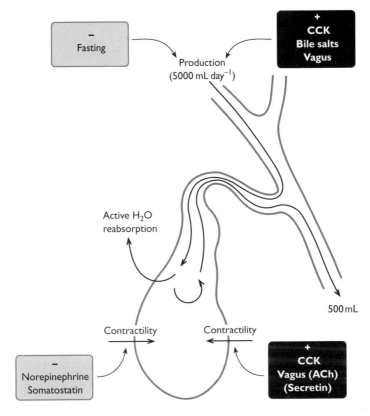

Fig. 61 Gallbladder function and activators

Gastro-oesophageal competence and gastro-oesophageal reflux disease

The gastro-oesophageal junction remains proof against return (reflux) of gastric contents into the oesophagus in health, preventing erosion of the relatively unprotected oesophageal mucosa. Several factors contribute to competence although none is exclusively important.

Intraluminal (gastro-oesophageal junction) pressure varies with time:
- minute-to-minute small variations (5–10 mmHg) due to lower oesophageal sphincter contraction;
- larger fluctuations (up to 80 mmHg) due to migrating motor complex (three per minute);
- contraction of crural diaphragm during inspiration (from 10 mmHg up to 150 mmHg during deep inspiration).

Factors maintaining competence and their disorders

Physiological

1 *Lower oesophageal 'sphincter' tone*: resting contractile tone in the circular smooth muscle of the lower oesophagus. Increase in response to gastrin, CCK, motilin and vagal stimulation (ACh). Reduced response to progestogens and oesophageal myenteric reflex.
   - Tone lost in pregnancy, and after vagal denervation and forcible dilatation of the oesophagus.
   - Operation of fundoplication aims to replace this with an external 'cuff' of stomach to provide contractile tone.

2 *Intra-abdominal oesophagus*: 2-cm length of oesophagus lying within the abdomen allows positive intra-abdominal pressure to keep the lower oesophagus closed below the area of negative intra-thoracic pressure.
   - Effect is lost in hiatus hernia where the gastro-oesophageal junction slides up into the chest. Operations to reduce gastro-oesophageal reflux are widely thought more likely to succeed if the gastro-oesophageal junction is reduced into the abdomen again.

Anatomical

1 *Crural sling*: striated muscle fibres from the right crus of the diaphragm slinging around the lower oesophagus act to close the lumen and support the smooth-muscle sphincter zone.
- Lost in obesity and hiatus hernia.
- Crural tightening is often added to other operations to reduce reflux.

2 *Acute angle (of His) between stomach and oesophagus*: acts to accentuate effects of sphincter and crural sling.
- Lost in obesity and hiatus hernia.
- Operation of gastropexy aims to recreate this angle.

3 *Mucosal rosette*: thickened and heaped mucosa at gastro-oesophageal junction helps to form a low-pressure seal.
- Lost in scarring due to acid reflux and corrosive ingestion.

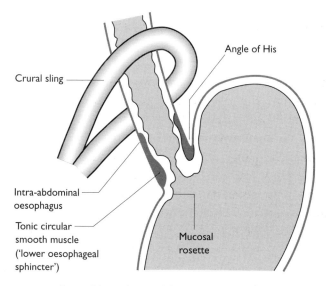

Fig. 62 Gastro-oesophageal junction and factors influencing competence

Duodenal functions and duodenectomy

Function	Normal	Postduodenectomy
Control of gastric acid chyme	The duodenal mucosa is relatively more resistant to gastric acid attack than more distal small-bowel mucosa and provides a location where neutralization can occur. Neutralization is provided in part by pancreatic bicarbonate secretion but also by bicarbonate output from the glands of Brunner in the duodenal submucosa. Glands of Brunner are stimulated directly by acid present in the lumen and by secretin	Rerouting of gastric contents risks ulceration of the recipient small-bowel mucosa and potential incomplete chyme neutralization with associated Fe^{2+}, Ca^{2+} and PO_4^- malabsorption as well as impaired fat emulsification
Gastric acid output stimulation	Duodenal G (gastrin) cells provide continued (late or intestinal phase of) gastric acid secretion stimulation	Of little importance
Gastric acid output reduction and reduced gastric emptying	Duodenal S (secretin) cells and M (GIP-secreting) cells act to reduce gastrin output from antral G cells as well as reducing gastric emptying and increasing pyloric tone	Loss of gastric acid and emptying control may contribute to dumping syndrome (p. 111) due to uncontrolled passage of gastric contents into small bowel
Pancreatic activation	Secretin from the duodenum causes alkaline fluid production by pancreatic ductal cells and promotes enzymatic secretion. CCK from the duodenum causes enzymatic secretion and promotes alkaline fluid production	
Bile production and flow	CCK and to a lesser extent secretin promote bile formation and precipitate biliary tree and particularly GB contraction	Loss of pancreatic and biliary stimulation causes a transient reduction in digestive capability particularly for fats and may lead to fat intolerance

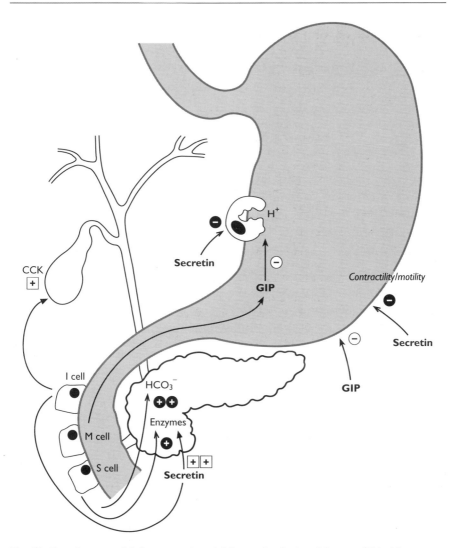

Fig. 63 Duodenum with hormonal activities and relationship to gallbladder, stomach and pancreas

Effects of terminal ilectomy

Removal of the terminal ileum (terminal ilectomy) may be accompanied by several disorders of selective reuptake of substances performed by the terminal ileum.

Bile-salt reuptake

- Minority occurs passively by transmembrane diffusion.
- Majority occurs via receptor-mediated Na^+ cotransport on the apical membrane of mucosal cells in the terminal ileum (final 60–80 cm).
- Reuptake 90% efficient.
- Loss of reuptake causes:
    (a) presence of bile salt in colonic contents which alters colonic bacterial growth and stool consistency;
    (b) increased bile-salt manufacture by the liver by cholesterol transmetabolism. Promotes cholesterol gallstone formation (so-called 'lithogenic bile').

Vitamin B_{12} uptake

- Occurs via receptor-mediated uptake across apical membrane of mucosa.
- Requires intrinsic factor (see p. 110) as cofactor to allow receptor binding.
- Intrinsic factor is liberated back into intestinal lumen once the vitamin B_{12} is bound onto the receptor.
- Loss of uptake causes deficient intake and progressive depletion of long-term vitamin B_{12} stores. Resultant B_{12} deficiency causes macrocytic anaemia with neurological degeneration of spinal cord white matter.

Water reabsorption

- Occurs through the lower ileum but particularly in the elderly the terminal ileum plays an important role in reducing the volume of fluid stool delivered to the colon.
- Loss of this capacity may increase stool frequency in the elderly.

γ-Globulin uptake

Reuptake of secreted γ-globulin may play a role in immune surveillance through the profuse lymphatic tissue in the wall of the terminal ileum.

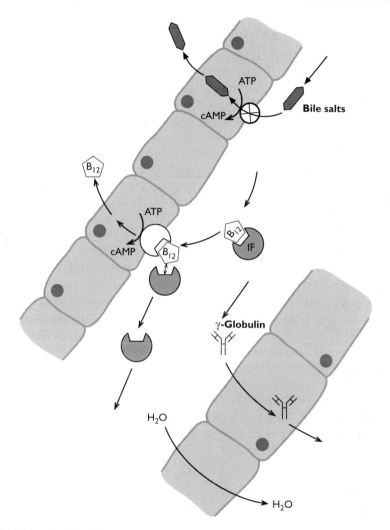

Fig. 64 Terminal ileal functions
IF, intrinsic factor.

The pancreas and consequences of pancreatectomy

Exocrine secretion

- Average output 1.4 L day^{-1}.
- Composition alters with increased flow rate, with pH rising.
- Contains proenzymes (trypsinogen, chymotrypsinogen, proelastase, pro-collagenase, procarboxypeptidases) for protein digestion.
- Contains proenzymes (prolipase, prophospholipase, colipase, α amylase) for fat digestion.
- Contains HCO_3, which alkalinizes duodenal and jejunal contents so aiding Fe^{2+}, phosphate and Ca^{2+} absorption in the upper small bowel.
- Stimulated by CCK (enzymes particularly), secretin (inorganic ions particularly), vagal tone (ACh), VIP and direct duodenal stretch reflexes.
- Suppressed by glucagon and somatostatin.

Endocrine secretion

- Occurs from islet cells found mainly in the body and tail.
- Includes insulin (from β cells), glucagon (α cells), somatostatin (δ cells) and pancreatic polypeptide (PP).

Effects of total pancreatectomy

- Loss of proteolytic proenzymes results in inadequate protein digestion and eventual protein malnutrition. Daily nitrogen balance becomes negative and progressive weight loss ensues.
- Loss of lipolytic proenzymes results in inadequate fat digestion. This exacerbates calorie malnutrition and results in a fatty chyme which promotes colonic bacterial growth causing a fat-laden, offensive, low-density stool with prominent flatus due to colonic bacterial activity.
- Loss of fat emulsification and digestion causes associated fat-soluble vitamin malabsorption (vitamins A, D, E, K).
- Loss of alkalinization of the chyme causes Fe^{2+}, Ca^{2+} and phosphate malabsorption with negative ion balance and eventual iron deficiency and osteoporosis.
- Loss of endocrine glucose regulation leads to diabetes mellitus of particularly difficult nature due to total loss of intrinsic insulin output *and*

counterregulatory glucagon which normally helps to correct the effects of overadministration of exogenous insulin.

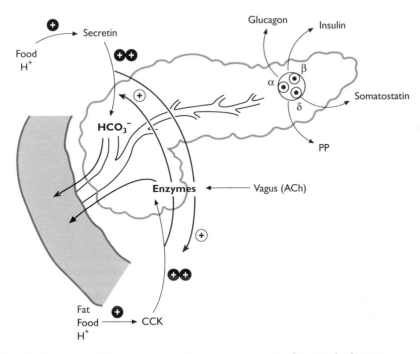

Fig. 65 Pancreas with hormone and enzyme output and controls shown

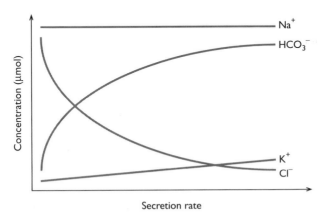

Fig. 66 Pancreatic secretion composition vs. secretion rate

Digestion and absorption

Protein digestion and absorption

Site	Enzyme (proenzyme)	Substrate	End product
Stomach	Pepsin(ogen)	Proteins	Polypeptides
Exocrine pancreas	Trypsin(ogen)	Proteins	Polypeptides
	Chymotrypsin(ogen)	Proteins, polypeptides	Polypeptides, oligopeptides
	Carboxypeptidases A, B	Polypeptides, oligopeptides	C-terminal amino acids, peptide residues
Enterocyte brush border (membrane bound)	Oligopeptidases	Polypeptides, oligopeptides	N- or C-terminal amino acids
	Aminopeptidases	Tripeptides, dipeptides	Amino acids
Cytosol	Dipeptidases	Dipeptides	Amino acids

- Dipeptides are absorbed more rapidly than amino acids by a distinct mechanism.
- Amino acid transport mechanisms (mostly Na^+ cotransport mechanisms) are specific for:
    - (a) neutral amino acids (tryptophan, alanine) (defective in Hartnup disease);
    - (b) dibasic amino acids (arginine, ornithine, lysine) and cystine (neutral) (defective in cystinuria);
    - (c) imino acids (proline, hydroxyproline) and glycine;
    - (d) dicarboxylic acids (glutamic and aspartic acids).

Carbohydrate digestion and absorption

Site	Enzyme	Substrate	End product
Salivary glands	Amylase	Starch	Oligosaccharides
Exocrine pancreas	Amylase	Starch	Oligosaccharides
Enterocyte brush border	Amylase	Starch, oligosaccharides	Maltose, glucose
	Sucrase	Saccharose, sucrose	Fructose, glucose
	Maltase	Maltose	Glucose
	Lactase	Lactose	Galactose, glucose

Virtually all carbohydrates are absorbed as monosaccharides. Glucose and galactose are transported by a Na^+ (secondary active) cotransport mechanism. Glucose enters the interstitium by simple or facilitated diffusion. Fructose is transported by facilitated diffusion into the enterocyte and largely converted to glucose. Basolateral Na^+ extrusion by Na^+/K^+-ATPase is the primary active process.

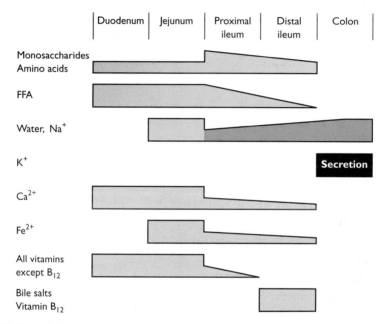

Fig. 67 Bowel absorption
FFA, free fatty acids.

Lipid metabolism

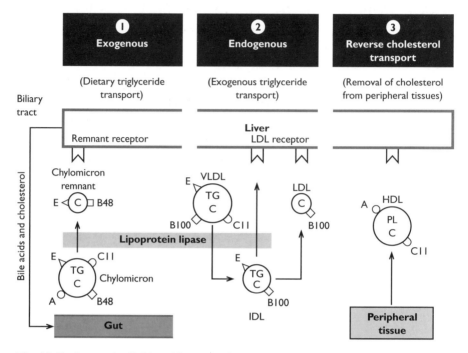

Fig. 68 Pathways for lipid and lipoprotein metabolism

In the exogenous pathway (1), chylomicrons transporting dietary lipids away from the gut are hydrolysed by lipoprotein lipase to chylomicron remnants. These are taken up by the liver. In the endogenous pathway (2), VLDLs are hydrolysed by lipoprotein lipase to give IDLs which are either taken up by the liver (via LDL receptors) or converted to LDL (also taken up by liver and other tissues via LDL receptors).

Reverse cholesterol transport (3) refers to the removal of cholesterol from peripheral tissues by HDLs. HDLs may be converted to IDLs or LDLs by lecithin cholesterol acyltransferase before uptake by liver.

Lipids: C, cholesterol; PL, phospholipid; TG, triglycerides. Apoproteins: A, B48, B100, E. Lipoproteins: HDL, high-density lipoprotein; IDL, intermediate-density lipoprotein; LDL, low-density lipoprotein; VLDL, very low-density lipoprotein.

Primary hyperlipidaemias are associated with mutations of:
- LDL receptor (familial hypercholesterolaemia);
- apoprotein B100 (familial hypercholesterolaemia);
- apoprotein E (remnant particle disease);
- apoprotein C2 (chylomicronaemia); and
- lipoprotein lipase (chylomicronaemia).

Polygenic hypercholesterolaemia and familial combined hyperlipidaemia (also primary hyperlipidaemias) are probably a result of multiple genetic and environmental factors.

5

Hepatic physiology

Total blood flow to the artery

= 350 mL min^{-1} via hepatic artery + 1100 mL min^{-1} via portal vein

= 1450 mL min^{-1} (29% cardiac output).

Functions of liver

Metabolic

- Carbohydrate metabolism:
  - (a) primary site of gluconeogenesis;
  - (b) glycogen storage;
  - (c) converts galactose and fructose to glucose.
- Fat metabolism:
  - (a) β oxidation of fats to acetyl coenzyme A (which enters Kreb's cycle);
  - (b) cholesterol, phospholipid and lipoprotein synthesis;
  - (c) fat synthesis from carbohydrates and proteins.
- Protein metabolism:
  - (a) protein synthesis;
  - (b) amino acid deamination;
  - (c) formation of urea.

Storage

- Vitamins D, A, B$_{12}$.
- Iron (in the form of ferritin).

Coagulation

- Prothrombin, clotting factors V, VII, IX and X (to some extent VIII), fibrinogen, plasminogen and antithrombin III synthesis. (Vitamin K is essential for the synthesis of prothrombin, protein C, and factors VII, IX and X.)

Endocrine

- Hormone metabolism (oestrogen, cortisol, aldosterone): usually production of inactive metabolites conjugated to glucuronic acid and excreted by kidney.

• Target organ of many hormones and production of circulating second messengers (insulin-like growth factors (IGFs)).

Other

• Drug metabolism.
• Formation and excretion of bile.
• Important site of haematopoiesis in fetal life.

Classification of jaundice

	Prehepatic	Hepatic	Extrahepatic
Test			
Bilirubin (3–17 μmol L⁻¹)	50–150	50–250	100–500
AST (< 35 iu)	< 35	300–3000	35–400
ALP (< 250 iu)	< 250	< 250–700	> 500
GGT (15–40 iu)	15–40	15–200	80–600
Albumin (40–50 g L⁻¹)	40–50	20–50	30–50
Reticulocytes (< 1%)	10–30	< 1	< 1
Causes			
Common	Neonatal	Cirrhosis	Common duct stones
		Hepatitis (viral, alcoholic)	Pancreatic cancer
		Hepatic metastases	
		PBC	
Uncommon	Haemolysis	Drug induced	Sclerosing cholangitis
	Gilbert's syndrome	Leptospirosis	Pancreatitis
		Liver abscess	
Rare	Crigler–Najjar	Cardiac failure	Portal lymphadenopathy
	syndrome	Wilson's disease	

AST, aspartate aminotransferase; ALP, alkaline phosphatase; γ-GT, γ-glutamyl transferase; PBC, primary biliary cirrhosis.
(Adapted from Travis, S.P.L., Taylor, R.H. & Misiewicz, J.J. (1998) *Gastroenterology*, 2nd edn. Blackwell Science, Oxford.)

5

| Enterohepatic circulation | | Causes of jaundice |

❶ Haemolysis

❷ Gilbert's syndrome

❸ Crigler–Najjar syndrome

❹ Dubin–Johnson syndrome

❺ Cholestasis: intra- or extrahepatic

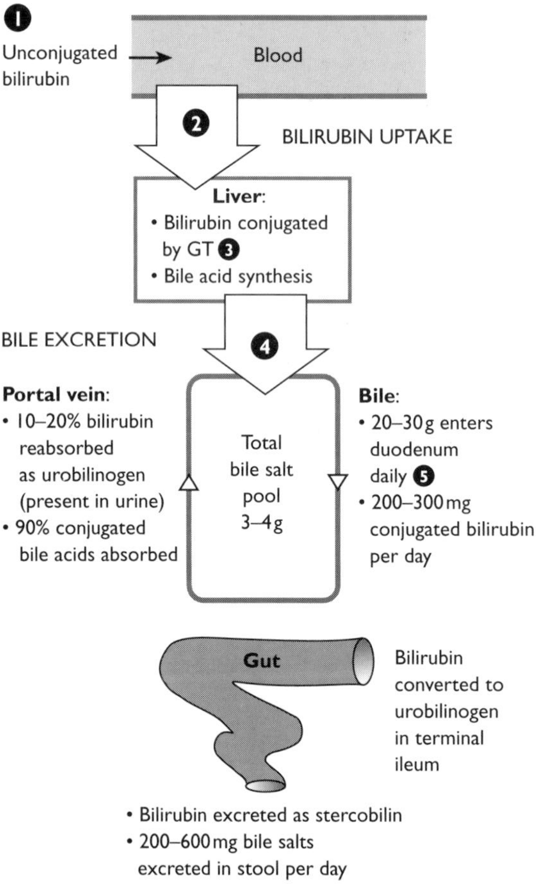

❶ Unconjugated bilirubin → **Blood**

❷ BILIRUBIN UPTAKE

Liver:
- Bilirubin conjugated by GT ❸
- Bile acid synthesis

BILE EXCRETION

❹

Portal vein:
- 10–20% bilirubin reabsorbed as urobilinogen (present in urine)
- 90% conjugated bile acids absorbed

Total bile salt pool 3–4 g

Bile:
- 20–30 g enters duodenum daily ❺
- 200–300 mg conjugated bilirubin per day

Gut

Bilirubin converted to urobilinogen in terminal ileum

- Bilirubin excreted as stercobilin
- 200–600 mg bile salts excreted in stool per day

Fig. 69
GT, glucuronyl transferase.

6: ENDOCRINE PHYSIOLOGY

6

Thyroid physiology

Actions of Triiodothyronine (T_3)

Overall, T_3 acts mainly by binding to nuclear receptors which alter the rate and balance of mRNA synthesis, so altering the spectrum of cell protein production. Few actions are mediated by direct cell receptor activation.

Calorigenesis

- Acts on all tissues except brain, lymphoid tissues and gonads.
- Onset of activity of single 'dose' is delayed by 6–12 h and action persists for 24–48 h.
- Maximum effect of single 'dose' is inversely proportional to catecholamine level.
- Probable mechanisms of action are:
  - (a) to facilitate sympathetic-mediated increase in cellular metabolism;
  - (b) direct effect on cell metabolism to uncouple oxidative phosphorylation.

Catecholamine synergy

- Facilitates β-receptor activation in normal receptors.
- Increases β-receptor production.
- Amplifies β-receptor activation response.

This causes widespread β effects systemically, particularly cardiac (see p. 178), and also activates β-receptor functioning in the ascending reticular activating formation of the brainstem raising levels of function of many aspects of cerebral activity.

Growth and development

Thyroxine (T_4) is important before and after birth for normal myelination and axonal and inter-cell development. T_3 becomes important in early childhood and facilitates growth hormone (GH) activity of cells.

Permissive effects

T_3 acts permissively on many pathways by its protein synthesis control. T_3 is five times more active than T_4.

- Promotes protein and fat catabolism with reduction in cholesterol production.
- Promotes N loss and negative nitrogen balance.
- Promotes cellular carbohydrate uptake.
- Promotes erythropoiesis.
- Promotes vitamin A production from absorbed carotene.
- Promotes bone mineralization.
- Permits neuronal functioning.

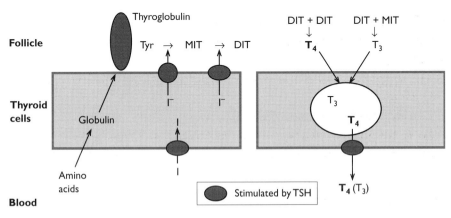

Fig. 70 Production of T₃ and T₄
85% of T_3 from peripheral conversion of T_4. 95% T_3/T_4 protein bound. DIT, di-iodotyrosine; MIT, mono-iodotyrosine; TSH, thyroid-stimulating hormone (thyrotrophin).

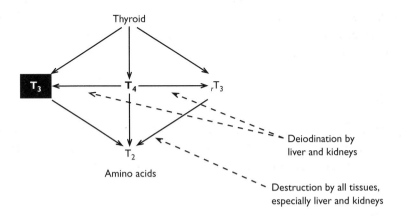

Fig. 71 T₃ and T₄ metabolism
Relative tissue activities: $T_3 = 5$; $T_4 = 1$; $_rT_3 = \rightarrow 0$.

Thyroid disorders

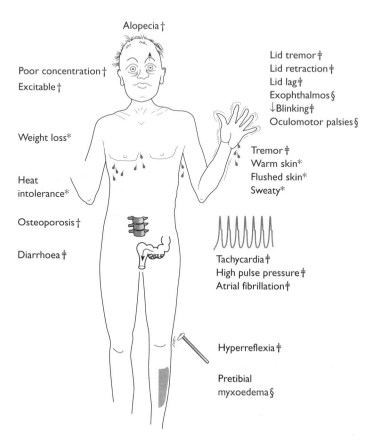

Alopecia †

Poor concentration †
Excitable †

Lid tremor ‡
Lid retraction ‡
Lid lag ‡
Exophthalmos §
↓Blinking ‡
Oculomotor palsies §

Weight loss *

Heat
intolerance *

Tremor ‡
Warm skin *
Flushed skin *
Sweaty *

Osteoporosis †

Diarrhoea ‡

Tachycardia †
High pulse pressure †
Atrial fibrillation †

Hyperreflexia ‡

Pretibial
myxoedema §

Fig. 72 Clinical features of hyperthyroidism
* Related to calorigenesis.
† Related to permissive effects.
‡ Related to catecholamine synergy.
§ Related to growth disorder.

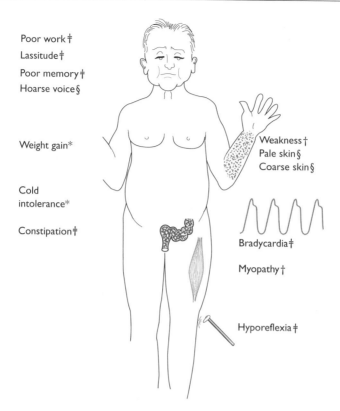

Poor work†
Lassitude†
Poor memory†
Hoarse voice§

Weight gain*

Cold
intolerance*

Constipation†

Weakness†
Pale skin§
Coarse skin§

Bradycardia‡

Myopathy†

Hyporeflexia‡

Fig. 73 Clinical features of hypothyroidism
* Related to calorigenesis.
† Related to permissive effects.
‡ Related to catecholamine synergy.
§ Related to growth disorder.

Interpretation of thyroid function tests (TFTs)

	Free T_4	Free T_3	TSH
Hyperthyroidism			
Graves' disease	↑	↑	↓
Toxic nodular goitre	↑ or N	N or ↑	↓
Pituitary hyperthyroidism	↑	↑	↑
Hypothyroidism			
1° hypothyroidism	↓	Unreliable	↑
2° (pituitary) or 3° (hypothalamic) hypothyroidism	↓	–	↓ (or N because abnormal TSH reacts with assay)
Subclinical hypothyroidism	N	–	↑

Adrenal cortex physiology and disorders

Actions of cortisol

Cortisol acts by binding to intranuclear receptors which regulate the transcription of specific genes.

Permissive homeostasis

- Permits vascular smooth muscle reactivity to α and β catecholamines.
- Permits muscle resistance to fatigue, particularly cardiac muscle.
- Permits mineralocorticoid (aldosterone) action on renal tubules.
- Permits antidiuretic hormone (ADH) activity on renal collecting duct (see p. 84).
- Permits gluconeogenesis.
- Permits GH action on cells to facilitate growth.
- May have a permissive role in the maintenance of normal body temperature through augmenting the role of T_3 in thermogenesis (see p. 130).
- Role in parturition.

Bone-marrow effects

- Allows bone-marrow response to increase in neutrophil, red cell and platelet production.
- Reduces T-cell number and function.
- Reduces B-cell clonal expansion.
- Reduces eosinophil and basophil counts.

Stress response (to surgery or trauma)

- Agonism of catecholamine effects.
- Promotes protein catabolism.
- Promotes lipolysis (permissive to GH and catecholamines).
- Increases neuronal excitability.

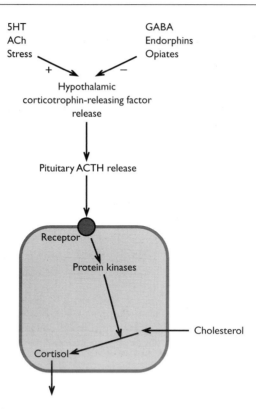

Fig. 74 Control of cortisol synthesis and release
ACh, acetylcholine; ACTH, adrenocorticotrophic hormone; GABA, γ-aminobutyric acid; 5HT, 5-hydroxytryptamine.

Relative activities of adrenal hormones

	Mineralocorticoid	Glucocorticoid
Corticosterone	15	0.3
Cortisol	1	1
Aldosterone	3000	0.3

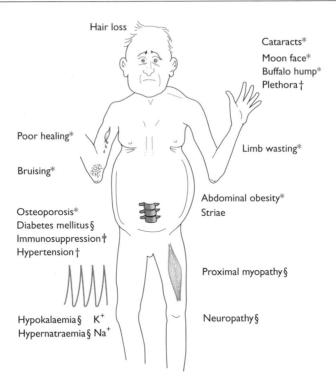

Fig. 75 Clinical effects of hyperglucocorticoidism
(usually known as Cushing's syndrome)

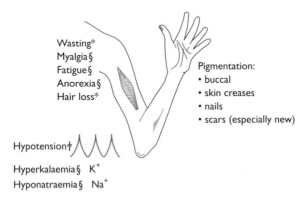

Fig. 76 Clinical effects of hypoglucocorticoidism
(usually known as Addison's disease)
* Effects due to alteration of protein/fat metabolism in stress response.
† Effects due to effects on vascular smooth muscle.
‡ Effects due to blood and other stress response effects.
§ Effects due to alteration of 'endocrine' status.

Calcium balance

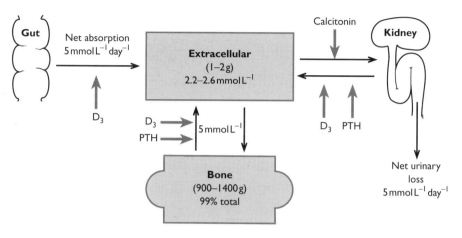

Fig. 77 Calcium fluxes
PTH, parathyroid hormone.

Parathyroid hormone

- Eighty-four amino-acid polypeptide produced in parathyroid glands.
- Produced from prepro-PTH by proteolysis to pre-PTH and PTH.
- Secretion is induced by hypocalcaemia and hyperphosphataemia. Ca^{2+}-sensitive receptors are thought to control the setpoint for Ca^{2+} sensing in the parathyroid gland.
- PTH-related peptide (PTHrP) is locally produced in bone and smooth muscle cells (SMC) and has a predominantly autocrine/paracrine effect. Plasma concentrations are low except in lactating women.
- Both PTH and PTHrP act via a common receptor expressed predominantly in bone, growth plate cartilage and kidney.
- Secretion of PTH altered within 30 min of change in serum Ca^{2+} level.

Actions (raises serum Ca^{2+}, lowers serum PO_4^{2-})
- Causes shrinkage of osteoblast cells on bone mineralized surface allowing osteoclasts access to osteoid so allowing bone resorption and Ca^{2+} release.
- Increases renal tubular binding of Ca^{2+} so increasing reabsorption.
- Decreases renal tubular binding of PO_4^{2-} so decreasing reabsorption.
- Increases 1α-hydroxylase activity so increasing vitamin D_3 production.
- SMC relaxation.

137

Vitamin D$_3$ (1,25-dihydroxycholecalciferol)

• Precursor (cholecalciferol) formed from ultraviolet light cleavage of 7-dehydrocholesterol in the skin or absorbed from dairy products in the ileum.
• Activated by 25-hydroxylation in the liver then 1α-hydroxylation in the kidney.
• Inactivated by 24-hydroxylation in the kidney.

Actions (raises serum Ca^{2+}, raises serum PO$_4^{2-}$)
• Increases ileal membrane cell absorption of Ca^{2+} by:
  (a) increasing apical membrane-binding protein;
  (b) increasing intracellular binding protein;
  (c) activating intracellular mitochondrial transport of Ca^{2+};
  (d) activating basal adenosine triphosphatase (ATPase)-dependent Ca^{2+} export from the cell.
• Increases tubular reabsorption of Ca^{2+} and PO$_4^{2-}$.

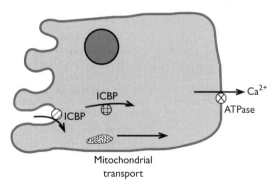

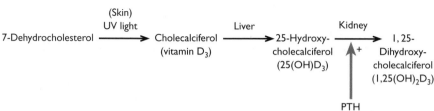

Fig. 78 Vitamin D$_3$ synthesis
ICBP, intracellular binding protein.

Calcitonin

- Thirty-two amino-acid polypeptide.
- Produced by interstitial ('C', 'chief') cells of thyroid gland.
- Produced from inactive precursor by a cleavage process activated by raised Ca^{2+} level.

Actions (lowers serum Ca^{2+}, lowers serum PO_4^{2-})
- Decreases tubular reabsorption of Ca^{2+} and PO_4^{2-}.
- Activates osteoblasts to promote mineralization of bone surface.

Hypercalcaemia	Hypocalcaemia
(a) Common	(a) Hypoparathyroidism
• Malignancy (55%)	• Idiopathic
• Hyperparathyroidism (35%)	• Acquired (hypomagnesaemia)
	• Pseudohypoparathyroidism
(b) Uncommon (<10%)	• Neonatal syndromes (Di George's syndrome)
• Renal dialysis + transplantation	(b) Renal insufficiency
• Thyrotoxicosis	
• Sarcoidosis	(c) Vitamin D deficiency
• Vitamin D excess	• Intestinal malabsorption
	• Vitamin D resistant rickets (hypophosphataemic)
(c) Rare (<1%)	• Fanconi syndrome
• Drugs (thiazides, lithium)	• Distal renal tubular acidosis
• Endocrine (Addison's acromegaly, phaeochromocytoma)	(d) Acute pancreatitis
	(e) Malignancy

6

- Ten grams per litre of albumin binds 0.2 mmol L^{-1} of Ca^{2+}. To correct for hypoalbuminaemia, add 0.1 mmol L^{-1} to total Ca^{2+} concentration for every 5 g L^{-1} decrease in albumin from the normal 40 g L^{-1}.
- Acidaemia decreases protein binding and increases ionized Ca^{2+}. For every 0.1 fall in pH, ionized Ca^{2+} rises by 0.05.

Magnesium balance

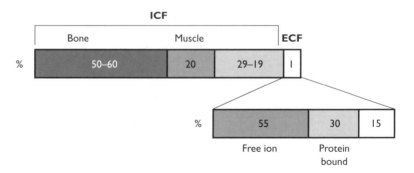

Fig. 79 Distribution
ECF, extracellular fluid; ICF, intracellular fluid.

One hundred milligrams of Mg^{2+} is excreted daily into the urine. Sixty to seventy per cent is reabsorbed via the thick ascending loop. Ten per cent is reabsorbed in the distal convoluted tubule (DCT) which is the major site of Mg^{2+} regulation. Hypomagnesaemia inhibits loop transport. Hypermagnesaemia promotes loop transport.

Functions

• Important cofactor of Na^+/K^+-ATPase (important for intracellular K^+ repletion).
• Promotes intracellular ATP restoration.
• Essential for adenylate cyclase activity (and therefore cAMP second messenger pathway).
• Activates Ca^{2+}-ATPase in sarcoplasmic reticulum (SR) which transports Ca^{2+} out of cytosol into SR (reducing SMC and cardiac muscle tone).
• Opposes catecholamine release.

Causes of hypermagnesaemia	Causes of hypomagnesaemia

Magnesium-containing enemas/laxatives/antacids
Burns, rhabdomyolysis
Hypothyroidism, diabetic ketoacidosis,
   mineralocorticoid deficiency
Respiratory acidosis
	Gastrointestinal losses:
	   Malnutrition
	   Malabsorption (small bowel)
	   Diarrhoea
	Renal losses:
	   Diuresis (diabetes, renal failure, postobstructive
	     nephropathy)
	   Renal tubular disorders
	   Hypercalcaemia
	   Drugs (diuretics, mannitol, aminoglycosides,
	     alcohol, cis-platinum)
	Other:
	   Acute pancreatitis
	   Conn's syndrome
	   Hyperthyroidism

6

Phosphorus balance

Fig. 80 Distribution

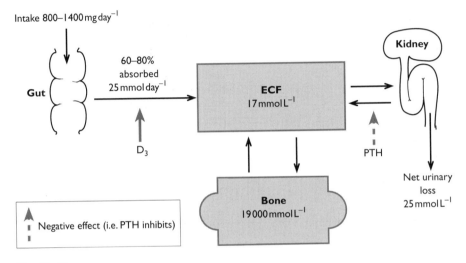

Fig. 81 Fluxes

- PO_4^{2-} is freely filtered in the kidney. Eighty per cent is reabsorbed in the proximal convoluted tubule (PCT) by Na–PO_4 cotransport (PTH sensitive).

Causes of hypophosphataemia	Causes of hyperphosphataemia
Decreased absorption:	Increased exogenous load:
Malnutrition	Oral supplementation
Vitamin D deficiency	Vitamin D intoxication
Phosphate binders	Phosphate-containing enemas
Chronic diarrhoea	Increased endogenous load:
Increased excretion:	Tumour lysis syndrome
Diuresis	Rhabdomyolysis
Hyperparathyroidism	Bowel infarction
Conn's syndrome	Haemolysis
Renal tubular disorders	Metabolic and respiratory acidosis
Alcohol abuse	Reduced excretion:
Metabolic acidosis	Renal failure
Intracellular shifts:	Hypoparathyroidism
Respiratory alkalosis	Acromegaly
Sepsis	Pseudohyperphosphataemia:
Drugs and hormones:	Multiple myeloma
Glucose	Hypertriglyceridaemia
Corticosteroids	
Insulin	
Glucagon	
Epinephrine	
Other:	
Hypokalaemia	
Acute gout	
Carcinoma	

6

Insulin

Produced by β cells in the pancreas.

Positive regulators	Negative regulators
Glucose	Sympathetic activity
Fatty acids and ketone bodies	Dopamine
Parasympathetic activity	Serotonin
Amino acids (arginine, leucine)	Somatostatin
Gut hormones (gastrin, cholecystokinin (CCK), secretin, glucagon, gastric inhibitory peptide (GIP))	
Prostaglandins	
Drugs (sulphonylureas)	

6

Actions

Metabolic effects (via insulin receptor on liver, fat and muscle cells):
- stimulates glucose transport and glycogen synthesis;
- stimulates lipogenesis;
- inhibits lipolysis.

Growth-promoting effects (via insulin-like growth factor 1 (IGF-1) receptor on muscle cells):
- stimulates cell growth and differentiation;
- stimulates DNA and RNA synthesis.

Mixed effects:
- stimulates amino acid influx into muscle and liver;
- stimulates protein synthesis;
- inhibits protein degradation.

Glucose balance

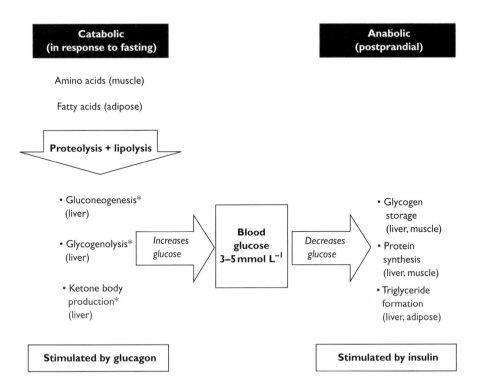

Fig. 82
* Inhibited by insulin.

Glucagon

Glucagon release is promoted by:
- catecholamines;
- sympathetic and parasympathetic stimulation;
- glucocorticoids;
- GI hormones (CCK, GIP);
- amino acids (alanine, arginine (promotes insulin release as well)).

Insulin-like growth factors (somatomedins)

IGF-1 and IGF-2 are produced by the liver in response to GH and modulated by binding to IGF-binding proteins.

IGF-1 binds to insulin and IGF-1 receptors and mediates the majority of growth-promoting (anabolic) effects of GH after birth.

Actions of IGF-1

- promotes local tissue and somatic growth;
- promotes cell metabolism;
- promotes cell survival (anti-apoptotic effect);
- amplifies local actions of trophic hormones (for example enhances gonadal steroid production and action).

IGF-2 acts via a mannose phosphate receptor and plays an important role in prenatal development.

Growth hormone

Growth hormone-releasing hormone (GHRH) is produced by the hypothalamus in response to central nervous system (CNS) neurotransmitters (catecholamines, serotonin, cholinergic substances) and acts on somatotroph cells in the pituitary gland via a single receptor. GHRH binding stimulates GH production via transcriptional proteins (which also activate prolactin and thyrotrophin). Somatostatin produced by the hypothalamus inhibits GH production via multiple receptors.

The GH family of genes on chromosome 17 includes:
- pituitary GH1;
- placental GH2 (encodes placental lactogen);
- chorionic somatotropic hormone genes (these appear to have no effect on growth).

Growth hormone receptor

- Growth hormone receptor (GHR) belongs to a cytokine receptor superfamily (includes receptors for prolactin, interleukins, erythropoietin, granulocyte colony-stimulating factor (GCSF) and granulocyte/macrophage colony-stimulating factor (GMCSF)) and activates tyrosine kinase.
- GH-binding protein found in serum is identical to the extracellular part of the GHR.
- Stimulates IGF-1 synthesis and secretion.
- Direct anabolic effects of GH via GHR are unclear.

Actions

Acute transient effects: hypoglycaemia, fall in free fatty acids (FFA).

Chronic effects: diabetogenic (hyperglycaemic); lipolytic (FFA rise, ketones rise); low-density lipoprotein (LDL) cholesterol falls; protein synthesis.

Causes of growth hormone deficiency
- Congenital (includes mutations of GH, Pit-1 and PROP-1 genes)
- Acquired: craniopharyngiomas, pituitary tumours, radiation
- Chronic renal insufficiency
- Turner's syndrome

Pregnancy

The changes of pregnancy

The typical changes in physiological parameters over a normal 40-week pregnancy given below occur due to the effects of various changes shown opposite.

Cardiovascular

- ↑ Blood volume by 30%.
- ↑ Cardiac output by 30%.
- ↓ Peripheral resistance by up to 30%.

Respiratory

- ↑ Alveolar ventilation rate.
- ↓ Mean $P_a\text{CO}_2$.
- Mild respiratory alkalosis.

Haematological

- ↓ Red blood cell count but ↑ red blood cell mass ⎫
- ↓ Haemoglobin (Hb) concentration but ↑ Hb mass ⎬ Dilution effect
- Normal cell size and corpuscular Hb.
- ↑ White cell count.
- ↑ Erythrocyte sedimentation rate (ESR).
- ↑ Coagulation proteins.

Endocrine

- ↑ T_4 production but ↑ thyroxine-binding globulin (TBG) production, thus normal free T_4 level.
- ↑ Cortisol production but ↑ cortisol-binding globulin (CBG) production, thus normal free cortisol.
- ↑ Prolactin.
- ↑ Insulin output but ↑ insulin resistance.

Gastrointestinal

- ↓ Transit time, thus ↑ constipation.
- ↓ Lower oesophageal sphincter pressure, thus ↑ reflux of gastric contents.

Renal

- ↑ Glomerular filtration rate.
- ↓ Renal threshold for excretion of glucose amongst other molecules.

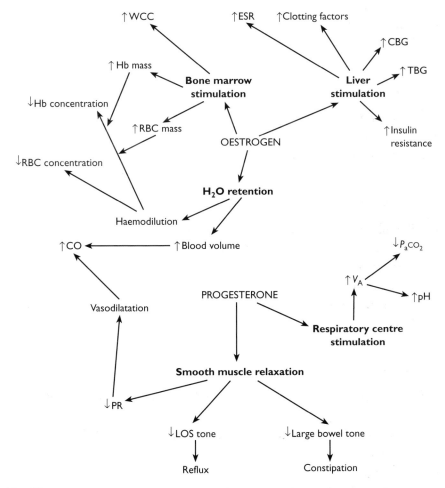

Fig. 83

LOS, lower oesophageal sphincter; PR, pulse rate; RBC, red blood cell; WCC, white cell count.

Effects of trauma/surgery

Net effects

The overall effects of the changes opposite are to give rise to a phase of metabolism called the *catabolic adrenergic–corticoid phase* which is characterized by the following.

Raised basal metabolic rate

This raises overall calorie requirement before the role of obligatory proteolysis is considered. Without adequate calorific intake proteolysis is promoted as a source of calories even in the absence of the hormonally derived effects listed below.

'Diabetogenic' state

The blood glucose usually remains normal except in profound sepsis or shock where there may be acute glycogen depletion due to consumption of glucose. If this consumption occurs sufficiently rapidly, before medium-term counterbalancing mechanisms have come into effect, hypoglycaemia can ensue.

Ketogenic state

Blood ketone levels are usually normal since increased production is met with increased metabolism and increased excretion in the urine. The presence of acid ketone bodies (acetoacetate and 3-hydroxybutyrate) makes this a potentially acidotic state.

Proteolytic state

Proteolysis results in a phase of acute negative nitrogen balance due to consumption of mainly muscle protein stores. This is an obligatory state. Adequate administration of calories can reduce but not prevent proteolysis mainly due to the actions of glucocorticoids. Severe stress leads to more widespread usage of proteins such as serum proteins particularly in conditions where there is impaired liver function. The increased production leads to raised urinary and plasma urea levels.

The weight loss seen in the response to trauma is a combination of proteolysis and ketolysis.

Hypernatraemic state

The net effects of the actions of ADH and aldosterone are not in complete balance and there is a tendency for sodium retention to exceed water retention although, unless this is confounded by additional administration of sodium exogenously such as NaCl infusions, serum levels are usually maintained.

Hormones that are increased

- ↑ ADH   →⎫ H_2O retention ++
- ↑ Aldosterone →⎬ Na^+ retention +++
             →⎭ K^+ loss +

	Gluconeogenesis	Ketolysis	Proteolysis	Negative N balance
Cortisol	++	+++	+++	++
GH	+	+		
Glucagon	+++	+	+	+
Epinephrine (adrenaline)	++	++		
Renin	↑ Aldosterone			

Hormones that are unaffected

- $T_4/T_3/TSH$.
- Follicle-stimulating hormone (FSH)/luteinizing hormone (LH).
- Insulin.

Associated findings

- ↑ Platelets   ⎫ ↑ Clotting tendency
- ↑ Fibrinogen ⎭
- ↑ Acute-phase proteins.

Arachidonic acid metabolism

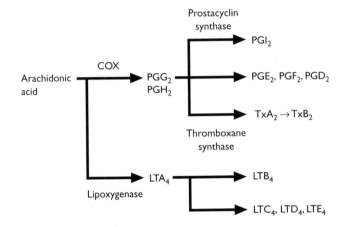

Fig. 84
COX, cyclooxygenase (can be either COX 1 which is constitutively active or COX 2 which is induced by cytokines and growth factors).

	Actions
Prostaglandins	
Prostacyclin (PGI_2)	Vasodilatation (particularly resistance vessels) by increasing K permeability and hyperpolarizing the cell as well as inhibition of endothelin production
	Inhibition of platelet aggregation
	Inhibition of cell migration and proliferation
	Inhibition of leukocyte adhesion
	Cytoprotective effects (possibly by scavenging reactive O_2 species)
PGE_1	Maintains patent ductus arteriosus
PGE_2	Suppresses interleukin 2 (IL-2) production by T cells
	Inhibits B-cell growth and differentiation
PGH_2	Vasoconstriction
Leukotrienes	
LTB_4	Enhances leukocyte migration and activation, neutrophil activation
LTC_4	Vasoconstriction
LTD_4	Increased microvascular permeability
LTE_4	Bronchoconstriction
Thromboxanes	
TXA_2	Vasoconstriction
	Platelet aggregation

Summary of pituitary hormones

Anterior pituitary

Cell	Hormone	Function
Corticotrophs: stimulated by CRF and vasopressin	ACTH	Stimulates glucocorticoid and androgen synthesis by adrenal cortex; maintains size of zona fasciculata and reticularis of cortex
Thyrotrophs	TSH	Stimulates production of thyroid hormones by thyroid follicular cells
Gonadotrophs	FSH	Stimulates ovarian follicle development; regulates spermatogenesis in testis
Gonadotrophs	LH	Causes ovulation and formation of corpus luteum in the ovary; stimulates oestrogen and progesterone production by the ovary; stimulates testosterone production by the testis
Lactotrophs	Prolactin	Stimulates secretion of milk
Somatotrophs	GH	(See p. 147)

CRF, chronic renal failure.

Posterior pituitary (neurohypophysis)

Region	Hormone	Function
Paraventricular nucleus	Oxytocin	Stimulates contraction of mammary myoepithelial cells during lactation; stimulates uterine contraction during labour
Supraoptic nucleus	ADH	(See p. 84)

7: PHYSIOLOGY OF THE BLOOD

7

Erythropoiesis

Steps involved in red cell production

1 Erythropoietin

- Produced by juxtaglomerular apparatus (JGA) of kidney in response to hypoxia.
- $t_{1/2} = 6$ h.
- Causes committed stem cells to activate and undergo DNA synthesis.
- Promotes speed of maturation of activated red cell precursors.
- Promotes early release of reticulocytes into circulation at high levels.

Deficiency: causes anaemia due to hypoproduction; caused by renal parenchymal damage (failure).

Excess: causes excess production (polycythaemia); caused by erythropoietin-secreting renal, endometrial, liver and neuroendocrine tumours as well as chronic hypoxia stimulating renal cells.

2 Cell division

- Occurs in response to local growth factors.
- Stimulated by circulating factors such as testosterone, cortisol and growth hormone.
- Requires vitamin B_{12}, folate, vitamin C, free T_3.

Deficiency: of vitamin B_{12} or folate causes fewer, larger cells with normal haemoglobin (Hb) level (macrocytic orthochromic anaemia); caused by dietary deficiency, poor uptake, excess utilization.

Deficiency: of testosterone etc. causes anaemia with normal red cells due to hypoproduction.

3 Cell membrane formation

Requires production of a lattice of specialized proteins to stabilize red cell membrane in biconcave form.

Abnormalities: cause deformed, fragile unstable red cells; caused by genetic abnormalities of protein production (e.g. spherocytosis, elliptocytosis, pyropoikilocytosis).

4 Iron uptake

• Occurs from circulating transferrin.
• Added to synthesized porphyrin ring to form haem in mitochondria.
Deficiency: causes small, pale red cells (microcytic hypochromic anaemia); caused by dietary deficiency, reduced absorption, chronic blood loss.

5 Haemoglobin synthesis

Occurs in two halves.
• **Globin chains** formed in cell cytoplasm—requires amino acids: adult forms α, β; infantile forms α, γ; fetal forms α, ε, ξ.
Deficiency: of globin causes microcytic, hypochromic anaemia; caused by abnormal globin genes reducing production (e.g. thalassaemia α or β).
• **Porphyrin ring** formed in mitochondrial space—requires pyridoxine, vitamin C, free T_3.
Deficiency: causes abnormal iron accumulation in red cells with low complete Hb (sideroblastic anaemia); caused by deficiency of pyridoxine, T_3 or vitamin C, or drugs.

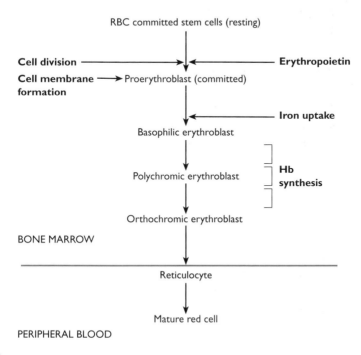

Fig. 85

Haemolysis

Features of red cell breakdown (haemolysis)

Abnormal red cells

Occur due to:
- cell membrane damage occurring with age, altering surface (glyco)proteins—average $t_{1/2}$ of red cell 120 days;
- cell membrane damage from drugs or chemicals;
- cell damage from mechanical injury;
- antibody attachment to cell surface due to autoimmune reaction;
- antibody attachment to cell surface of transfused blood cells;
- cell membrane damage from abnormal intracellular proteins.

    Mostly removed from circulation by spleen or liver. Mechanical damage, some autoantibodies and drug/chemical damage result in destruction within vessels (intravascular haemolysis).

Intravascular haemolysis

- Free intravascular Hb may be taken up by liver or spleen macrophages.
- Free intravascular Hb may be filtered at the glomerulus and excreted in the urine causing haemoglobinuria.

Extravascular haemolysis

- Occurs in splenic and liver macrophages.
- Haem is divided from globin and cleaved.
- Iron is recycled into circulating transferrin stores or extracellular haemosiderin stores.
- Porphyrin ring is converted to bilirubin and circulated to the liver attached to albumin or prealbumin. Bilirubin is taken up by surface receptors.
- Intracellular transport of bilirubin occurs attached to protein X or Y.
- Glucuronidation occurs in cytoplasm and glucuronidated bilirubin is excreted into bile by active transport for excretion.
- Globin chain is cleaved into constituent amino acids and recycled.

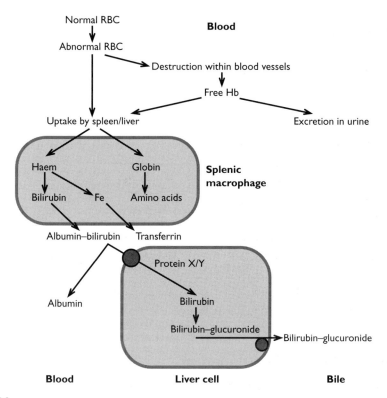

Fig. 86
RBC, red blood cell.

Causes of haemolytic anaemia

Intracorpuscular	Extracorpuscular	Other
• Membrane: hereditary spherocytosis, elliptocytosis • Defects in glycolysis: G6PD deficiency (may be secondary to drugs) • Chloramphenicol, sulphonamides, dapsone • Aspirin • Antimalarials • Pyruvate kinase deficiency • Haemoglobinopathies: thalassaemia, SCD • Other: Wilson's disease	• Autoimmune (Coombs' positive) (a) Isoantibodies: haemolytic disease of the newborn, ABO incompatibility (b) Autoantibodies: warm (SLE, lymphoma); cold (mycoplasma, EBV, lymphoma) • Drugs (penicillin, cephalosporins mefenamic acid, methyldopa) • Paroxysmal cold haemoglobinuria	• Infection (malaria, *Clostridium, Meningococcus*) • Hypersplenism • Microangiopathic (HUS, TTP) • Trauma (prosthetic valves) • Paroxysmal nocturnal haemoglobinuria

EBV, Epstein–Barr virus; G6PD, glucose-6-phosphate dehydrogenase; HUS, haemolytic–uraemic syndrome; SCD, sickle cell disease; SLE, systemic lupus erythematosus; TTP, thrombotic thrombocytic purpura.

Haemostasis and clotting

Phases of haemostasis

1 Vascular phase. Transmural pressure and flow of blood are reduced by:
  (a) vasospasm in response to direct mechanical effects;
  (b) vasospasm in response to locally released mediators;
  (c) local oedema increasing tissue pressure;
  (d) haematoma formation.

2 Platelet phase. Adherence of platelets to damaged endothelium activated by ADP and collagen, binding von Willebrand factor (vWF) to endothelium which binds to platelet receptors Ib. The integrin glycoprotein IIb/IIIa binds to fibrinogen and vWF and plays an important role in platelet aggregation and adhesion. Aggregation of platelets then occurs due to release of mediators (ADP, 5-hydroxytryptamine (5HT), platelet aggregating factor, thromboxane A_2) from activated platelets. Activated platelets also release platelet factor 3 (PF3).

3 Clotting phase. The division into two pathways is artificial since, *in vivo*, coactivation of the twin arms results in a unified process.

4 Monocytes. Activated monocytes express tissue factor and FV/Va. They also have high-affinity binding sites for FX/Xa and fibrinogen.

Central roles of thrombin in haemostasis

- Fibrin production.
- XIIIa production.
- Platelet aggregation.
- Platelet activation.
- VII, V activation.
- Plasmin activation (see p. 162).

Abnormalities in haemostasis

1 Vascular phase, e.g. vitamin C (collagen) deficiency resulting in poor activation of kallikrein and weak vessels.

2 Platelet phase, e.g.

(a) aspirin-induced reduction of thromboxane A_2 production reduces platelet activation and aggregation;

(b) haematological disorders (e.g. leukaemia) resulting in reduced platelet counts;

(c) autoantibodies resulting in impaired platelet activation and adherence or platelet destruction (idiopathic thrombocytopaenic purpura, ITP).

3 Clotting phase, e.g.

(a) inherited disorders of reduced clotting factor production (e.g. Christmas disease (IX), Hageman disease (XII), haemophilia (VIII));

(b) reduced production of liver factors by liver disease (e.g. cirrhosis);

(c) reduced production of factors II, VII, IX, X by warfarin treatment;

(d) reduced Ca^{2+} activity due to sequestration (e.g. citrate in blood transfusion storage).

4 Abnormalities in antithrombotic pathways (see p. 162).

7

Measures of haemostatic activity

- Vascular and platelet phase combined = Bleeding Time.
- Intrinsic pathway = Activated Partial Thromboplastin Time.
- Extrinsic pathway = Prothrombin Time.
- Common pathway = Thrombin Time.

Clotting regulation

Fibrinolysis

Plasmin

- Derived from plasminogen.
- Formed from two polypeptide chains.
- Produced by the action of thrombin or plasminogen activators (from ECs (tissue plasminogen activator, TPA) or within the plasma) on plasminogen.
- Serine protease in action.
- Attacks unstable bonds between fibrin molecules to generate fibrin degradation products.
- Activates the lysis of prekallikreins.
- Lyses V and VIII in small quantities.
- Inhibited by α_2 antiplasmin from the liver and plasminogen activator inhibitor 1 from ECs.

7

Antithrombin III

- Binds thrombin irreversibly with great avidity.
- Also binds XIIa, IXa and XIa to deactivate the molecules.

Protein C/S

- Activated by thrombin binding to endothelial thrombomodulin receptors.
- Binds Va and VIIIa to prevent thrombin generation.
- Activates fibrinolytic system.
- Needs Ca^{2+}.

Tissue factor pathway inhibitor

- Produced by platelets in response to thrombin and ECs.
- Inhibits factor Xa (FXa) and tissue factor (TF)/VIIa complex.

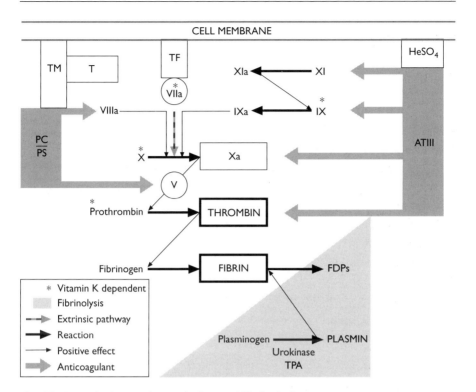

Fig. 87 Coagulation, anticoagulation and fibrinolysis
FDPs, fibrin degradation products; HeSO$_4$, heparin sulphate; PC and PS, protein C and
protein S; T, thrombin; TM, thrombomodulin.

Both low-molecular-weight heparins and unfractionated heparin bind to
antithrombin III (ATIII) to accelerate its interaction with Xa. Only
unfractionated heparin catalyses the inactivation of thrombin by ATIII.

Functions of the spleen and splenectomy

Functions of the spleen

Red pulp: haematological functions

• Sequestration and destruction of abnormal red cells, platelets and exogenous matter. Circulating red cells presenting abnormal surface or cytoplasmic antigens or proteins become attached to resident macrophages within the red pulp. Macrophage function disassembles the component parts of the red cell structure, recycling amino acids into the circulation, extracting ionic iron from Hb for storage/recirculation coupled to ferritin and producing bilirubin for plasma carriage attached to albumin. Abnormalities detected and removed include:

(a) loss of membrane structure (e.g. hereditary spherocytosis, hereditary elliptocytosis);

(b) abnormal membrane structure (e.g. drug-induced damage (Cu^{2+}), autoantibodies, malarial parasite infection, oxidized membrane proteins in G6PD deficiency);

(c) abnormal internal proteins (e.g. HbH $\{\beta_2\beta_2\}$ in thalassaemia A, HbS in sickle cell disease, HbM in methaemoglobinaemia, drug-induced Hb oxidation (sulphasalazine));

(d) abnormal platelets with altered surface antigens;

(e) miscellaneous debris (e.g. hyperlipidaemia, inorganic dyes, circulating microspheres).

• Storage of red blood cells. In humans, storage capacity is usually less than 50–100 mL.

White pulp: immunological functions

• Production of lymphocytes. Lymphocytic production is maximal in embryonic life and early infancy but remains even in adulthood with capacity for considerable expansion in the face of bone-marrow disorder.

• B-cell response. B lymphocytes in the spleen are responsible for a significant proportion of rapid immunoglobulin M (IgM) production in acute bacterial infections.

• T-cell response. T lymphocytic response in systemic cell-mediated immunity is partly mediated by T cells present in periarterial lymphatic sheath.

Effects of splenectomy

- Peripheral blood film.

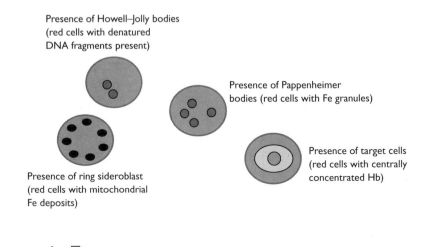

Presence of Howell–Jolly bodies (red cells with denatured DNA fragments present)

Presence of Pappenheimer bodies (red cells with Fe granules)

Presence of ring sideroblast (red cells with mitochondrial Fe deposits)

Presence of target cells (red cells with centrally concentrated Hb)

Increased platelet count

Fig. 88

- Loss of B-cell-mediated antibody production. Manifest as:
  (a) increased susceptibility to meningococcal and pneumococcal sepsis (organisms possessing capsules for which antibody is required for macrophage function);
  (b) overwhelming postsplenectomy sepsis syndrome—rare syndrome of rapidly overwhelming (usually Gram-positive) septicaemia associated with reduced IgM response.

8: PHYSIOLOGY OF THE NERVOUS SYSTEM

8

Organization of the nervous system

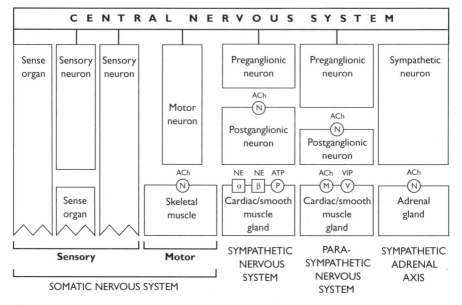

Fig. 89

ACh, acetylcholine; NE, norepinephrine; VIP, vasoactive intestinal peptide. Circles and squares represent receptors: α & β, adrenergic receptors; M, muscarinic; N, nicotinic.

Synaptic transmission

A synapse refers to any neuroeffector junction.

Neuromuscular junction

Each muscle cell is innervated by a single end plate.

Synaptic vesicle cycle

1 Neurotransmitter release:
   (a) synaptic vesicles filled with neurotransmitter dock in active zone;
   (b) fusion and exocytosis of synaptic vesicles in response to rapid Ca^{2+} influx during an action potential (AP);
   (c) neurotransmitter release and diffusion across synaptic cleft;
   (d) endocytosis of empty vesicles.
2 ACh within synaptic cleft by ACh esterase.
3 Binding to postsynaptic nicotinic ACh receptor (AChR) generates miniature end-plate potential (MEPP). The nicotinic AChR forms an ion channel which is closed in the resting state. Binding of ACh opens the channel transiently and allows the passage of cations.
4 Temporal and/or spatial summation of MEPPs generates large depolarization (end-plate potential (EPP)) which may reach the threshold for AP generation in a postsynaptic muscle cell.

Curare binds to AChR inducing paralysis. Botulinum toxin prevents ACh release presynaptically. α Bungarotoxin binds irreversibly to AChR.

8

Myasthenia gravis

Myasthenia gravis (MG) is characterized by weakness and fatiguability of skeletal muscle (extraocular, facial, proximal limb, diaphragm). The basic abnormality is a decrease in the number of nicotinic AChRs at the neuromuscular junction (NMJ) due to specific autoantibodies.

Neuroneuronal transmission

• Each neuron may be innervated by several thousand synapses.
• Synapses may be excitatory (bring postsynaptic cell closer to threshold for firing AP) or inhibitory. There is no equivalent of the inhibitory synapse at the NMJ.
• Cotransmission is the rule rather than the exception.

Important cotransmitters

• VIP is released with ACh at parasympathetic postganglionic synapses.
• ATP serves as a cotransmitter within the sympathetic nervous system. ATP is also released from endothelial cells and aggregating platelets. ATP has a dual action on blood vessels by its opposing actions at P2x purinoceptors on smooth muscle cells (SMCs) (vasoconstriction) and P2y purinoceptors on endothelial cells (ECs) (vasodilatation). Adenosine is generated by ATP degradation and has the following effects:

(a) inhibition of norepinephrine release from sympathetic nerve endings;
(b) vasodilatation via endothelium-dependent and -independent mechanisms;
(c) bronchoconstriction;
(d) antiarrhythmic properties (by interruption of atrioventricular (AV) node conduction).

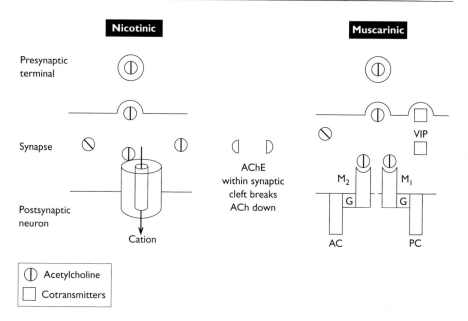

Fig. 90 Cholinergic synapses
AC, adenylyl cyclase; AChE, acetylcholinesterase; G, G protein; PC, phospholipase C.

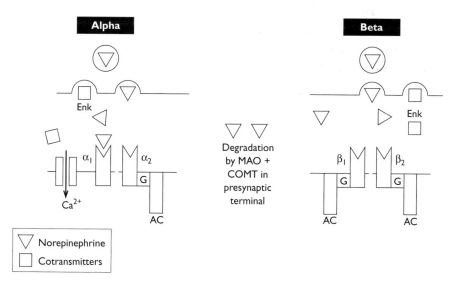

Fig. 91 Adrenergic synapses
COMT, catecholamine O-methyl transferase; Enk, enkephalin; MAO, monoamine oxidase.

Muscle physiology

Excitation–contraction coupling

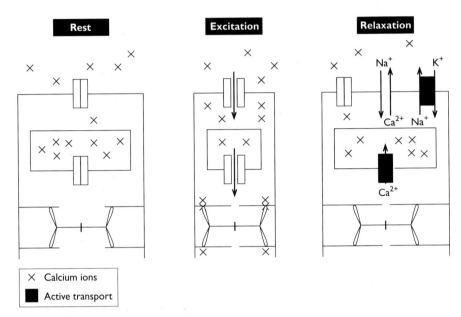

Fig. 92

Muscle cells are multinucleated, surrounded by membrane (sarcolemma) and contain several myofibrils surrounded by sarcoplasmic reticulum (SR) and invaginations of the sarcolemma known as transverse tubules (T system). The fundamental unit of muscular contraction is the myosin cross-bridge.

1 With electrical depolarization of the sarcolemma and T system, Ca^{2+} is released by the terminal cisterns of the SR into the cytoplasm.

2 Ca^{2+} binds the troponin complex (subunits C, I and T).

3 This displaces tropomyosin, uncovers myosin-binding sites on actin and facilitates actin–myosin interaction.

4 ATPase activity in the myosin head hydrolyses ATP to ADP.

5 This causes a conformational change in the myosin cross-bridge and thus sliding of myosin filaments relative to actin.

6 Myosin-binding protein C modulates contraction when phosphorylated.

7 During relaxation, troponin releases Ca^{2+}.

8 Ca^{2+} is pumped actively into the SR by a Ca^{2+}/Mg^{2+}-ATPase.

Muscle types

Skeletal muscle

	Type I	Type IIA	Type IIB
Myosin ATPase activity	Slow	Fast	Fast
Ca^{2+} pumping capacity of SR	Moderate	High	High
Diameter	Moderate	Small	Large
Glycolytic capacity	Moderate	High	High
Oxidative capacity (correlates with mitochondrial content, capillary density, myoglobin content)	High	High	Low
	Fatigue resistant	Fatigue resistant	Fatigue prone
Notes	Like cardiac muscle		Predominates in white muscle specialized for fine, skilled movement such as extraocular and some hand muscles

Smooth muscle

• Ultrastructural differences: fusiform, smaller, T tubules absent, SR poorly developed.

• Actin anchored to cell surface via intermediate filaments.

• Actin and myosin not arranged in a polarized fashion.

• Actin : myosin ratio variable (as high as 20 : 1).

• Rate of ATP hydrolysis 100-fold slower than with striated muscle; therefore capable of sustained contraction with minimal energy expenditure.

8

Reflexes

The reflex arc is the basic unit of integrated neural activity in which a particular stimulus elicits a stereotyped, specific response. A typical motor reflex will consist of a sense organ, an afferent neuron, one or more synapses (usually located in a ganglion or in the central nervous system (CNS)), an efferent neuron and an effector (such as a muscle).

Reflexes may be:
- monosynaptic (stretch reflex); or
- polysynaptic (inverse stretch reflex).

Alternatively, reflexes may be:
- axonal (involving peripheral unmyelinated nociceptive afferent nerve fibres only; scratching the skin, for example, stimulates an impulse which travels along a side branch of an afferent nerve, culminating in the release of substance P, histamine and local vasodilatation);
- spinal (see below); or
- central (involving the brainstem, for example; see p. 181).

Spinal reflexes

Postural reflexes	Righting reflexes	Protective reflexes
Stretch reflex	**Crossed extensor response**	**Withdrawal reflex**
Muscle contraction in response to stretch. Activation of sensory endings in the muscle spindle increases spindle discharge and produces muscle shortening via a negative-feedback mechanism. This serves to maintain muscle length, contributes to muscle tone and smooths movements	Extension of contralateral limb during withdrawal reflex	Painful stimulation of skin results in withdrawal of ipsilateral limb by contraction and relaxation of flexors and extensors respectively
		Abdominal reflex
Inverse stretch reflex		**Cremasteric reflex**
Strong muscle contraction and activation of the Golgi tendon organ result in muscle relaxation		
Extensor thrust		
In response to pressure on foot		
Plantar reflex		
Babinski sign		

Spinal cord injury

Causes

• Vertebral column injuries (particularly vertebral body burst injuries and fracture dislocations).
• Penetrating injuries (knife and bullet wounds).
• Spinal artery occlusion (thrombosis, aortic dissection).

Spinal shock

	Acute	Chronic
	(Normally lasts 10–21 days)	
	Fall in resting potential of neurons reduces response to normal reflex and background discharge	Normal resting potential returns Raised neuronal discharge due to hypersensitivity and localized upregulation by collaterals
Motor	Muscle groups flaccid Tendon and superficial reflexes absent	Muscles hypertonic Upper limb flexors predominate Lower limb extensors predominate Tendon reflexes increased in both speed and magnitude Superficial reflexes remain reduced
Autonomic	Hypofunction with vasodilatation and bradycardia (particularly in young individuals)	Autonomic functions are often controlled by new spinal 'control centres' below the lesion Activation of these centres may be precipitated by relatively minor stimulation of visceral receptors such as by coughing or the Valsalva manoeuvre Local spinal vasomotor centres restore a degree of control to blood pressure and vasoconstriction but with limited response
Bladder	Detrusor hypotonia and isolated (tonic) sphincter resulting in overflow incontinence	Resting detrusor tone with reflex contraction upon rising bladder volume Sphincteric tone maintained until reflex relaxation (producing reflex incontinence)
Rectum	Hypotonia with faecal retention	Lowered resting tone Occasional reflex mass emptying (often inadequate or absent)

8

Vision

Phototransduction

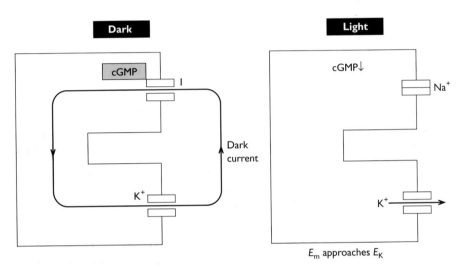

Fig. 93
E_m, membrane potential; E_K, potential due to potassium.

Phototransduction is the process by which light energy is converted to a change in the membrane potential of the photoreceptor cell (rod or cone).

The outer segment of the rod contains internal membranous discs which contain the light-sensitive protein rhodopsin. Rhodopsin consists of opsin bound to retinol.

In the dark, non-selective cation channels in the outer segment are bound to cGMP and open, causing a predominantly Na^+ influx. This is counterbalanced by an outward K^+ current in the inner segment. This is known as the dark current. The level of cGMP in the outer segment depends on the rate of synthesis (by guanylate cyclase) and degradation (by phosphodiesterases).

Absorption of a photon of light leads to isomerization of retinol, structural activation of rhodopsin (metarhodopsin formation) and activation of transducin.

Active transducin produces a fall in cGMP, closure of cGMP-gated cation channels and hyperpolarization of the rod (due to a continued K^+ efflux).

Eye movements

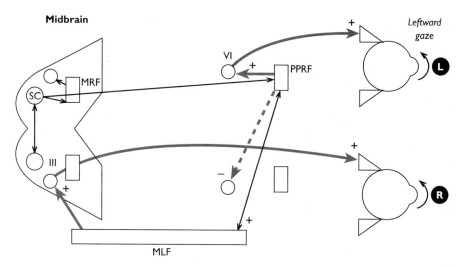

Fig. 94 Regulation of eye movements
MLF, medial longitudinal fasciculus; MRF, mesencephalic reticular formation; PPRF,
paramedial prepontine reticular formation; SC, superior colliculus; III, oculomotor nerve;
IV, abducent nerve.

Structures involved in eye movement

MLF	Receives fibres from various nuclei concerned with conjugate gaze
	A right MLF lesion results in failure of the right eye to adduct on leftward gaze
SC	Important in saccadic eye movements
	Role in sensory interaction between visual, auditory and somatosensory stimuli
PPRF	Responsible for horizontal saccades
	Receives input from cortex, cerebellum and SC; left PPRF stimulation produces leftward gaze

The cerebellum, basal ganglia and vestibular nuclei have important inputs
into the oculomotor system.

Autonomic nervous system

Organ	Sympathetic Effect	Receptor type	Parasympathetic (via muscarinic AChR) effect
Heart			
SA node	Tachycardia	β_1	Bradycardia
Inotropism	Positive	β_1	No effect
AV node	Increased conduction	β_1	AV block
Circulation			
Arterioles	Vasoconstriction	α	No effect
Veins	Venoconstriction	α	No effect
Viscera			
GI tract			
Smooth muscle	Reduced motility	α_2, β_2	Increased motility
Sphincters	Constriction	α_2, β_2	Relaxation
Pancreas	Decreased secretion	α	
Bladder			
Sphincter tone	Increased	α	
Detrusor tone	Decreased	β_2	
Bronchial smooth muscle	Bronchodilatation	β_2	Bronchospasm
Kidney	Renin secretion	β_2	
Eye			
Pupil	Dilatation	α	Constriction
Ciliary muscle	Relaxation	β	Contraction

GI, gastrointestinal; SA, sinoatrial.

Autonomic function testing

- Heart rate variation with deep breathing (respiratory sinus arrhythmia):
  (a) test of parasympathetic function;
  (b) normal variation ranges: 15–20 bpm (< 20 years old), 5–8 bpm (> 60 years old);
  (c) abolished by atropine.
- Valsalva response (see 'Cardiovascular physiology', p. 48).
- Orthostatic blood pressure recordings.
- Cold pressor test: immersion of hand in ice-cold water increases systolic blood pressure by 10–20 mmHg (spinothalamic afferent limb, sympathetic efferent limb).
- Sudomotor function (assessment of sweating).

8

Sympathectomy

Cervical sympathectomy above the T1 level

Effects (all ipsilateral) are:
- iris constriction (meiosis);
- reduced facial skin sweating (anhydrosis);
- increased facial skin blood flow (flushing);
- enophthalmos;
- loss of tone to levator palpebrae superioris (smooth muscle portion) (partial ptosis);
- nasal congestion.

Together known as Horner's syndrome, caused by total disruption to the origin of the cervical sympathetic chain supplying the ipsilateral head and neck vessels.

Causes are:
- trauma;
- local tumours;
- surgery (complication of).

Upper thoracic sympathectomy below the T1 level

Effects (all ipsilateral) are:
- reduced sweating in upper limb (anhydrosis);
- increased blood flow to upper limb skin (flushing).

Lower limb sympathectomy at lumbar level

Effects (all ipsilateral) are:
- reduced skin sweating (anhydrosis);
- increased blood flow to lower limb skin (flushing);
- reduced deep pain sensation;
- ischaemic rest pain (usually a byproduct of sympathectomy for ulceration);
- other deep pain.

May be performed by:
- open surgical approach;
- laparoscopic surgical approach;
- closed chemical approach either temporary using local anaesthesia or permanent using phenol.

8

Inotrope action and autonomic receptors

	Receptor type				HR	MAP	CO	PVR	Bronchodilatation	Renal blood flow
	β_1	β_2	α	Dopamine						
	Positive chronotropic and inotropic effect	Vasodilatation	Vasoconstriction	Renal, mesenteric, coronary and cerebral vasodilatation (via DA_1 receptors) and inhibition of NE release and peripheral vasodilatation (via DA_2 receptors)						
Dopamine	++	0	+ to ++	++	+/++	+	+++	+	0	+++
Dobutamine	++	+	+	0	+	+	+++	−	0	+
Epinephrine (adrenaline)	++	++	+ to ++	0	+++	+	+++	+/−	++	− −
Norepinephrine (noradrenaline)	++	0	++	0	−	+++	+/−	+++	0	− − −
Isoprenaline	++	++	0	0						

CO, cardiac output; DA_1, dopamine type 1; DA_2, dopamine type 2; HR, heart rate; MAP, mean arterial pressure; PVR, peripheral vascular resistance.

8

180

Brainstem reflexes, decerebration and decortication

Brainstem reflexes

- Pupillary reflexes:
  (a) direct light reflex;
  (b) consensual light reflex.
- Vestibulo-ocular reflexes.
  (a) **Cephalo-ocular** (oculocephalic) reflexes. Doll's eye movements are reflex movements elicited by movement of the head from side to side. Eye movements are in the opposite direction to head movement. Full movements exclude a brainstem lesion. In the normal awake subject, movements are suppressed by cerebral influences to allow visual fixation. In coma due to bihemispherical damage, eye movements are loose or easier due to disinhibition of the brainstem reflexes.
  (b) Caloric testing. Irrigation of the external ear with cool water causes a convection current within the endolymph of the inner ear. Tonic deviation of the eyes to the side of cool water irrigation indicates an intact brainstem pathway.
- Facial nerve (grimacing) reflex.
- Corneal reflex. Normal bilateral eye closure in response to corneal stimulation. The reflex is lost if reflex pathways between the fifth and seventh cranial nerves within the pons are lost.
- Gag reflex.

8

Criteria for brain death

- Pupils fixed and unresponsive to light.
- Absent corneal reflexes.
- Absent pain response in cranial nerve distribution.
- Absent gag reflex on endotracheal tube movement.
- Oculocephalic reflexes absent (absence of doll's eye response).
- Vestibulo-ocular reflexes absent (no nystagmus).
- No spontaneous respiration after 10 min (patient ventilated on 100% O_2 at 4 breaths min^{-1} and tidal volume of 7 mL kg^{-1}).
  (Other criteria include: core body temperature > 35°C, no CNS-depressant drugs for > 48 h, no neuromuscular blocking drugs for > 12 h.)

Decerebration

Transection of the brainstem at the level of the midbrain (between superior and inferior colliculi, above respiratory centres) produces overactivity of excitatory neural centres. Lesions below the lateral vestibular nucleus or above the red nucleus do not result in decerebration.

Causes

- Few traumatic injuries cause such a localized division of the neuraxis.
- Rupture of basilar artery aneurysm with subarachnoid haemorrhage.
- Tumour.

Pathophysiology

- Loss of positive output from the red nucleus (midbrain) to inhibitory reticular formation (hindbrain). The rubrospinal tract (emerging from the red nucleus) predominantly stimulates α and γ flexor motor neurons.
- Unopposed positive output of lateral vestibular (Deiter's) nucleus, resulting in excitation of α and γ extensor motor neurons in the cord.

Effects

- Decerebrate rigidity with all four limbs held in the maximally extended position (elbows and wrists extended with arm pronation).
- Generalized hyperreflexia.
- Absence of descending inhibition reveals primitive pontine and medullary mediated postural reflexes:
    (a) tonic labyrinthine reflex (reduces the degree of ipsilateral extensor tone with positioning of the head on that side);
    (b) tonic neck reflex (increases the degree of contralateral extensor tone with the act of turning the neck to that side);
    (c) grasping reflex (complex grasping action upon stimulation of the palmar skin).

Decortication

Division of the neuraxis above the level of the thalamus and hypothalamus can occur due to:

- severe ischaemic or hypoxic injury to upper cortical centres;
- bilateral internal capsule haemorrhage or infarction.

Descending cortical modulation of brainstem reflexes is lost. Thalamic reflexes and hypothalamic visceral integration are maintained.

Decortication is also associated with generalized hypertonia, hyper-reflexia (although much less than in decerebration) and mild spasticity of lower limb extensor groups and upper limb flexor groups (elbow and wrist flexion with arm supination).

Blood–brain barrier and cerebral blood flow

Blood–brain barrier

Components

- Tight junctions between ECs lining cerebral capillaries.
- Astrocyte foot processes enveloping the cerebral capillaries.

Effects

- Formed in the third month of gestation but only fully active by the age of 7 years.
- Proteins near totally excluded (very slow penetration) (e.g. organic dyes).
- Polar molecules highly excluded (e.g. penicillin).
- Amino acids and glucose, carrier-mediated uptake.
- Close control of ion flow across the barrier.
- Lipid-soluble substances freely permeable (e.g. anaesthetic agents, opiates).

Sites excluded from the barrier under normal circumstances

- Neurohypophysis (posterior pituitary).
- Tuber cinerum (subfornical organ).
- Supraoptic crest.
- Area potrema (chemotactic trigger zone).

Causes of breakdown of the barrier

- Trauma.
- Radiation damage.
- Ischaemia.
- Infection.
- Inflammation.
- Tumours.

Cerebral blood flow

Figures

- Blood flow: 50 mL per 100 g of tissue min^{-1}.
- O_2 delivery: 3.3 mL per 100 g of tissue min^{-1}.
- Weight 1.5 kg (2% body weight)
  - ∴ 50 mL min^{-1} O_2 consumption (18% of total body consumption)
    750 mL min^{-1} blood flow (15% of cardiac output).
- Ratio of blood flow grey : white matter = 6 : 1

Effectors

- Local autoregulation is the main determinant under normal conditions. Method of control is not clear. Likely to be a mixture of smooth muscle tonicity (myogenic) and responses to local metabolites (e.g. pH, K^+, etc.).
- Major systemic effector is P_aCO_2.
- Little role for vasomotor nerves (little effect during hypotensive vasoconstriction).

Responses in pathology

- Absence of O_2 delivery:
  - (a) 5–10 s—unconciousness;
  - (b) 5 min—neuronal death (cortical neurons most susceptible, brainstem neurons least susceptible);
  - (c) 7–10 min—brainstem death.
- Causes of loss of normal autoregulation:
  - (a) global—blunt, concussive trauma, intracranial haemorrhage;
  - (b) local—tumours, infarction.

8

Physiology of cerebrospinal fluid

Formation

- 0.5 mL min^{-1}.
- 60% by choroid plexuses in lateral ventricles and 40% by exposed vessels on the inner surface of the ventricular walls.
- Active process involving carbonic anhydrase (inhibited by acetazolamide).

Composition

- 150 mL average.
- Clear, colourless.
- No red cells, no neutrophils, < 5 monocytes mL^{-1}.
- Protein 20 mg 100 mL^{-1} = 0.2 g L^{-1}.
- Glucose 50% of plasma.
- Cl$^-$, CO$_2$, creatinine higher than plasma.
- Ca^{2+}, cholesterol, urea lower than plasma.

Circulation

- Pressure 10–13 cmH$_2$O (lying down).
- Free equilibration with subarachnoid and brain extracellular fluid (ECF).

Functions

- Supports brain tissue.
- Protects brain tissue against deforming forces.
- Regulates the brain ECF.
- Provides protein transport and control of levels.

Reabsorption

- Depends on cerebrospinal fluid (CSF) pressure (zero at < 70 mmH$_2$O, 0.5 mL min^{-1} at 115 mmH$_2$O).
- Occurs via:
  (a) the arachnoid villi (80%);
  (b) the spinal rootlets (15%);

(c) the olfactory membrane;
(d) the dura mater.

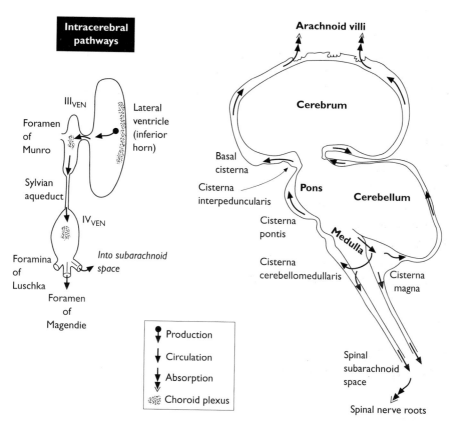

Fig. 95

9: APPENDIX

9

Statistics

Errors

- Type I: incorrect rejection of the null hypothesis when it is actually true (an apparent difference where none exists).
- Type II: incorrect retention of the null hypothesis when it is false (no apparent difference when there is one)—caused by small sample sizes.

Reliability

The likelihood of a test giving the same result when conducted by two different observers or on two different occasions.

Validity

The assessment of test results against an absolutely correct standard (e.g. screening tests against pathological diagnoses).

	Actual result	
Test result	Positive	Negative
Positive	a	b
Negative	c	d

Sensitivity

$$\frac{\text{No. true positives}}{\text{No. true positives} + \text{No. false negatives}} \quad \frac{a}{a+c}$$

(the ability of the test to correctly include positive results)

Specificity

$$\frac{\text{No. true negatives}}{\text{No. true negatives} + \text{No. false positives}} \quad \frac{d}{d+b}$$

(the ability of the test to correctly exclude negative results)

Positive predictive value

$$\frac{\text{No. true positives}}{\text{No. true positives} + \text{No. false positives}} \qquad \frac{a}{a+b}$$

(the chance that a positive result is correct = the probability of a given patient's positive result being a correct result)

Standardized mortality

The corrected mortality rate for a population if that population had the same experience and composition as a standard population:

$$\text{SMR}_A = \frac{\text{Death rate}_A}{\text{Death rate}_{ST}} \times \frac{\text{Population}_{ST}}{\text{Population}_A} \times 100$$

where A is the study population and ST is the standard population.

9

Amino acids

Essential (non-synthesizable by the body)

- Phenylalanine.
- Valine.
- Tryptophan.
- Leucine.
- Isoleucine.
- Lysine.
- Methionine.
- Threonine.

9

Plasma proteins

Functions

Transport

CO_2, O_2, urea, creatinine, glucose, fatty acids, lipids, amino acids, trace metals, enzymes (e.g. pseudocholinesterase), hormones, heat.

Defence

- Humoral: acute phase proteins, immunoglobulins.
- Cellular: non-specific (polymorphs), specific (lymphocytes).

Homeostasis

- Buffering.
- Oncotic pressure generation.
- Clotting and fibrinolysis.
- Heat control.
- Fluid volume preservation.

Types

Group	Concentration (g L^{-1})	MW	Function
α_1	1–5	5 000	Carriage, acute phase reactants
α_2	4–10	50 000	Carriage
β_1	6–10	90 000	Carriage
β_2		90 000	Carriage
γ	5–15	156 000	Humoral immunity
Albumin	35–50	69 000	Oncotic pressure, non-specific carriage
Fibrinogen	1.6–4.2	340 000	Clotting

MW, molecular weight.

Albumin

- 50% intravascular.
- 50% stored in skin and connective tissue.
- 10% recycled per day.

Normal parameters: biochemical

Serum levels

Na$^+$	135–145 mmol L^{-1}
K$^+$	3.5–5.0 mmol L^{-1}
Ca^{2+}	2.2–2.65 mmol L^{-1}
Mg^{2+}	0.7–1.0 mmol L^{-1}
Cl$^-$	95–115 mmol L^{-1}
PO$_4^{2-}$	0.65–1.45 mmol L^{-1}
HCO$_3^-$	24–33 mmol L^{-1}
Protein	41 g dL^{-1}
Urea	2.5–10 mmol L^{-1}
Creatinine	80–150 µmol L^{-1}
Bilirubin	7–20 µmol L^{-1}
Glucose	3.0–5.5 mmol L^{-1} fasting
	< 10 mmol L^{-1} random

Enzymes

Amylase	< 300 iu L^{-1}
Alkaline phosphatase	30–300 iu L^{-1}
AST	5–43 iu L^{-1}
γ-GT	7–51 iu L^{-1}
5NT	< 9 iu L^{-1}
Acid phosphatase	1–5 iu L^{-1}
LDH	240–530 iu L^{-1}
CK (% MB)	< 200 iu (6%)

Hormones

ACTH	10–80 ng L^{-1}
Aldosterone	< 500 pmol L^{-1}
Angiotensin II	5–35 pmol L^{-1}
ADH	0.9–4.5 pmol L^{-1}
Total T$_3$	1.2–3.0 pmol L^{-1}
Total T$_4$	50–141 nmol L^{-1}
TSH	< 6 mU L^{-1}
TBG	7–20 mg L^{-1}
FSH	< 5.0 U L^{-1}
LH	3–8 U L^{-1}
Prolactin	< 400 U L^{-1}
am Cortisol	160–700 nmol L^{-1}

AST, aspartate aminotransferase; γ-GT, γ-glutamyl transferase; 5HT, 5-hydroxytryptamine; LDH, lactate dehydrogenase; CK, creatine kinase; MB, myocardial type; ACTH, adrenocorticotrophic hormone; ADH, antidiuretic hormone; TSH, thyroid-stimulating hormone (thyrotrophin); TBG, thyroxine-binding globulin; FSH, follicle-stimulating hormone; LH, luteinizing hormone.

9

Normal parameters: haematological

Haematology

Hb	15.5/14.0 g dL^{-1}
PCV	0.47/0.43 %
RBCC	5.5/4.8 × 10^9 L^{-1}
MCV	78–100 fL
MCH	27–32 pg
MCHC	30–35 g dL^{-1}
WCC	4.0–11.0 × 10^9 L^{-1}
neutrophils	2.0–7.5 × 10^9 L^{-1}
lymphocytes	1.5–4.0 × 10^9 L^{-1}
monocytes	0.2–0.4 × 10^9 L^{-1}
eosinophils	0.04–0.4 × 10^9 L^{-1}
basophils	0.02–0.1 × 10^9 L^{-1}
platelets	100–400
ESR	< 20 mm h^{-1}

Mean values ♂/♀ (for Hb, PCV, RBCC)

Clotting times

Bleeding	3–6 min
Clotting	4–7 min
Prothrombin	10–15 s
Thrombin	11–15 s

Haematinics

Iron	11–31 µmol L^{-1}
TIBC	45–72 µmol L^{-1}
B$_{12}$	150–850 ng L^{-1}
Fibrinogen	1.6–4.2 g L^{-1}
Folate	3–18 µg L^{-1}
Ferritin	20–300 µg L^{-1}
FDPs	< 0.8 µg L^{-1}

9

FDPs, fibrin degradation products; Hb, haemoglobin; MCH, mean corpuscular haemoglobin; MCHH, mean corpuscular haemoglobin; MCV, mean corpuscular volume; PCV, packed cell volume; RBCC, red blood cell count; TIBC, total iron-binding capacity; WCC, white cell count.

Normal parameters: cardiovascular

		Mean
Cardiovascular		
CVP/RAP	2–10 mmHg	1–8 mmHg
RVP		
Systolic	15–30 mmHg	
Diastolic	0–8 mmHg	
PAP		
Systolic	15–30 mmHg	9–19 mmHg
Diastolic	4–12 mmHg	
PCWP/LAP	4–16 mmHg	2–12 mmHg
LVP		
Systolic	100–140 mmHg	
Diastolic	0–10 mmHg	
Aorta		
Systolic	100–140 mmHg	70–105 mmHg
Diastolic	60–90 mmHg	
Resting values		
Cardiac index	$2.8–4.2 \text{ L min}^{-1} \text{ m}^{-2}$	
Stroke volume	60–120 mL	
Heart rate	50–80 bpm	
Ejection fraction	0.52–0.67	
SVR	$800–1200 \text{ dyne s}^{-1} \text{ cm}^{-5}$	
PVR	$50–150 \text{ dyne s}^{-1} \text{ cm}^{-5}$	
Other		
BMI	19–26	

BMI, body mass index; CVP, central venous pressure; LAP, left arterial pressure; LVP, left ventricular pressure; PAP, pulmonary artery pressure; PCWP, pulmonary capillary wedge pressure; PVR, pulmonary vascular resistance; RAP, right atrial pressure; RV, right ventricular (pressure); SVR, systemic vascular resistance.

Normal parameters: respiratory

	Approx. in male 70 kg, 1.82 m
Typical values	
Tidal volume	500 mL
Forced vital capacity	5500 mL
Residual volume	1500 mL
Inspiratory reserve volume	4000 mL
Expiratory reserve volume	1000 mL
Functional residual volume	2500 mL
Total lung capacity	7000 mL
FEV_1	4400 mL
FEV_1/FVC	0.80
Blood gases	
P_aO_2	13.5 kPa
P_vO_2	5.3 kPa
P_aCO_2	5.3 kPa
P_vCO_2	6.1 kPa
O_2 consumption	250 mL min^{-1}
CO_2 production	200 mL min^{-1}
Respiratory quotient	0.8

FEV_1, forced expiratory volume in 1 s; FVC, forced vital capacity.

9

Recommended daily values

Nutrient	Recommended daily allowance (RDA)	Source
Vitamin A (β carotene)		Animal foods Dairy products Plant carotenes
Vitamin B_1 (thiamine)	1.35 mg	Cereals Liver
Vitamin B_2 (riboflavin)	1.5 mg	Dairy products
Vitamin B_6 (pyridoxine)	2.0 mg	Cereals Liver
Vitamin B_7 (niacin)	18 mg	Yeasts
Vitamin B_{12} (cyanocobalamin)	1.0 μg	Dairy products
Folate	0.2 mg	Green vegetables
Vitamin C		Fruit Vegetables
Vitamin D_3	5.2 μg	Dairy products
Vitamin E		Nuts/fish/wheat oils
Vitamin K		Liver Vegetables Cheese
Zinc	10–20 mg	
Copper	10–20 mg	
Iron	15 mg	

9